Volumes in This Series

Reports

1. *A Survey of Sardis and the Major Monuments outside the City Walls,* by George M. A. Hanfmann and Jane C. Waldbaum (1975)
2. *Sculpture from Sardis: The Finds through 1975,* by George M. A. Hanfmann and Nancy H. Ramage (1978)

Monographs

1. *Byzantine Coins,* by George E. Bates (1971)
2. *Ancient Literary Sources on Sardis,* by John G. Pedley (1972)
3. *Neue epichorische Schriftzeugnisse aus Sardis,* by Roberto Gusmani (1975)
4. *Byzantine and Turkish Sardis,* by Clive Foss (1976)
5. *Lydian Houses and Architectural Terracottas,* by Andrew Ramage (1978)
6. *Ancient and Byzantine Glass from Sardis,* by Axel von Saldern (1980), published jointly as A Corning Museum of Glass Monograph

ARCHAEOLOGICAL EXPLORATION
OF SARDIS

Fogg Art Museum of Harvard University
Cornell University
The Corning Museum of Glass
Sponsored by the American Schools
of Oriental Research

General Editors

George M. A. Hanfmann
Jane Ayer Scott

Monograph 6

Jointly published as
A CORNING MUSEUM OF GLASS MONOGRAPH

ANCIENT AND BYZANTINE GLASS FROM SARDIS

Axel von Saldern

Harvard University Press
Cambridge, Massachusetts
London, England   1980

*Publication of this volume was made possible by grants from
the Corning Glass Works Foundation
and the Samuel H. Kress Foundation*

WITHDRAWN

Library of Congress Cataloging in Publication Data

Saldern, Axel von, 1923–
    Ancient and Byzantine glass from Sardis.

    (Monograph—Archaeological Exploration of
Sardis ; 6)
    Bibliography: p.
    Includes index.
    1.   Glassware, Ancient—Turkey—Sardis.
2.   Glassware, Byzantine—Turkey—Sardis.
3.   Glassware—Turkey—Sardis.   4.   Sardis—
Antiquities.   I. Title.   II. Series:
Archaeological Exploration of Sardis.
Monograph ; 6.
NK5107.S2   748.299562   79-27959
ISBN 0-674-03303-5

CONTENTS

EDITORS' PREFACE

The Archaeological Exploration of Sardis began its work in 1958 as a joint effort of Harvard and Cornell Universities, under the general sponsorship of the American Schools of Oriental Research. This, the eighth volume in the Sardis series of final publications, is devoted to the very considerable finds of ancient and Byzantine glass made at Sardis from 1958 through 1978. We are pleased that this volume will appear simultaneously as A Corning Museum of Glass Monograph. The Corning Museum of Glass became a participant in the Archaeological Exploration of Sardis in 1960, and the entire program has benefited from a series of grants from the museum and the Corning Glass Works Foundation. We wish to thank the trustees, especially Arthur A. Houghton, Jr., and the successive directors of the museum, Paul N. Perrot, Robert H. Brill, and Thomas S. Buechner, for their assistance and interest in the proper study and publication of the glass finds excavated at Sardis.

We take this opportunity to express our profound gratitude to the government of the Republic of Turkey for the privilege of working at Sardis. The Department of Antiquities and Museums, formerly under the Ministry of the Prime Minister and now under the Ministry of Culture, and the Directors General, their officers and representatives, have been unfailing in their help. We honor here the memory of the late Director General, Hikmet Gurçay, who did so much to assist the work at Sardis and to further cooperation in the researches of Turkish and American scholars. The most complete and interesting glasses are at the Archaeological Museum in Manisa and the study material is stored in the depots of the expedition camp at Sardis. We owe a special debt of thanks to the successive directors and staff of the Manisa Museum, especially to its present director, Kubilây Nayır.

Both the excavation and research programs have been made possible by grants and contributions extending over two decades from the Bollingen Foundation (1959–1965), the Old Dominion Foundation (1966–1968), the Loeb Classical Library Foundation (1965–1970), the Wenner Gren Foundation for Anthropological Work (1967), the Charles E. Merrill Trust (1973), the Ford Foundation (1968–1972), and the Billy Rose Foundation (from 1970). Donations were received through the American Schools of Oriental Research, and Cornell University contributed university funds from 1957 through 1968. Much of the Harvard contribution came from the group of Supporters of Sardis, established in 1957, which includes both individuals and foundations. We owe the continuity of our work to their enthusiasm and generosity, and particularly to the advice and support of James R. Cherry, Landon T. Clay, Catharine S. Detweiler, John B. Elliott, Mrs. George C. Keiser, Thomas B. Lemann, Nanette B. Rodney, Norbert Schimmel, and Edwin Weisl, Jr.

The excavation of the sectors which yielded the glass finds and the stratigraphic contexts on which this study is based was aided by a grant in Turkish currency made by the Department of State to the President and Fellows of Harvard College for the years 1962–1965.[1]

1. No. SCC 29 543, under the Mutual Educational and Cultural Act, Public Law 87-256, and Agricultural Trade Development and Assistance Act, Public Law 480 as amended.

The National Endowment for the Humanities, through a series of grants largely on a matching basis, has played a key role in sustaining the Sardis program.[2] This vital help is most gratefully acknowledged. Our special gratitude goes to the friends and foundations who enabled the project to receive the Endowment's support through their matching contributions. In accordance with a request of the Endowment, we state that the findings and conclusions here do not necessarily represent the views of the Endowment.

In connection with this volume, the Sardis program was fortunate to have Axel von Saldern place his unrivaled knowledge of ancient and medieval glass at the disposal of the expedition. He agreed to undertake the study of glass from Sardis during his tenure as curator of The Corning Museum of Glass and completed it in spite of growing administrative responsibilities as curator of the Kunstmuseum Düsseldorf and as director of the Museum für Kunst und Gewerbe, Hamburg. We are greatly beholden to him for carrying through this task. At the Sardis research office at Harvard University, Electra D. Yorsz edited the preliminary manuscript and did much to clarify the organization and content. Ann B. Brownlee continued the editing and prepared the footnotes and illustrations. The plans and drawings were prepared for publication by Elizabeth Wahle and the map of glass sites (Plan I) is her original work. The photographs are largely the work of Elizabeth Gombosi. Debra A. Hudak put the book into final form and saw it through the press.

This volume was originally intended to include the results of a technical study undertaken by Robert H. Brill, Research Scientist, The Corning Museum of Glass. These researches will more appropriately ap-

pear within a major study, still in preparation, which will present analyses of some twelve hundred glasses from all periods and geographic locations. Scholars interested in the preliminary analytical results from the Sardis examples are encouraged to contact Dr. Brill at The Corning Museum of Glass, Corning, New York.

Our colleague in the general editorship, Stephen W. Jacobs, died suddenly on August 8, 1978. As Cornell University's representative, he had shaped the policies of the Sardis Expedition, especially in architectural research, and had inspired the great restoration program of the Marble Court of the Gymnasium at Sardis. We mourn the loss of this sage counselor and staunch friend.

Monograph 6 of the Sardis series represents studies of the material found at Sardis by excavations and research extending over twenty years and is the result of a broadly based effort. However, without the publication grants of the Corning Glass Works Foundation and the Samuel H. Kress Foundation this study could never have been published. We are most grateful to Dr. Franklin D. Murphy and Mary M. Davis of the Kress Foundation and to Thomas S. Buechner and Richard B. Bessey of Corning and to the boards of trustees for this generous assistance.

George M. A. Hanfmann
Jane Ayer Scott
Harvard University

2. Grant nos. H67-0-56, H68-0-61, H69-0-23, RO-111-70-3966, RO-4999-71-171, RO-6435-72-264, RO-8359-73-217, RO-10405-74-319, RO-23511-76-541.

AUTHOR'S PREFACE

The glass published here comes from the excavations undertaken between 1958 and 1978 by the Archaeological Exploration of Sardis of Harvard and Cornell Universities under the direction of George M. A. Hanfmann until 1976 and of Crawford H. Greenewalt, Jr., thereafter. It is with deep gratitude that I thank George Hanfmann for his invitation to study the finds from this site. His knowledge, kindness, and high professional standards are admired by his colleagues and students; I would like to join the chorus in praise of him and his wife, Ilse. I am, furthermore, most grateful to Jane Ayer Scott, whose never-ending patience was severely tested by my long delays in delivering the manuscript.

We were fortunate enough to have had the assistance of a dedicated staff who excelled in editing, correlating, typing, and proofreading the texts, catalogue entries, and statistical tables. My heartfelt thanks are due to Ann B. Brownlee, M. Elizabeth Craig, Debra A. Hudak, Christopher Ratté, Monika Salter, Ruth Weil, and Electra D. Yorsz. The excellent drawings are the work of Elaine Gazda and Elizabeth Wahle. Most of the photographs were taken by Elizabeth Gombosi.

I would like to thank the following persons and institutions for information, photographs, and permission to reproduce and publish illustrations: Kristin Anderson, Department of Classical Art, Museum of Fine Arts, Boston; Dan Barag, Hebrew University, Jerusalem; Mary Crawford, The Egypt Exploration Society; Sidney M. Goldstein, The Corning Museum of Glass; Clasina Isings, Rijksuniversiteit Utrecht; Philip J. King, President, American Schools of Oriental Research; Margrit Ludwig, Rheinisches Landesmuseum, Bonn; D. C. van den Oever, Wolters-Noordhoff bv, Groningen; Erwin Oppenländer, Waiblingen; Helmut Ricke, Kunstmuseum der Stadt Düsseldorf; Louise A. Shier, Curator, Kelsey Museum of Ancient and Mediaeval Archaeology, University of Michigan; and Elisabeth Stommel, Römisch-Germanisches Museum, Cologne.

The Corning Glass Works Foundation and The Corning Museum of Glass sponsored and financed four visits to Sardis in 1960, 1968, 1969, and 1978. I would like to extend my thanks to those authorities who were responsible for allocating the necessary funds. I join the editors in thanking Corning and the Samuel H. Kress Foundation for making the preparation and publication of this volume possible.

Axel von Saldern

Hamburg
November 1979

Technical Abbreviations

a.s.l.	above sea level	m.	meters
C (preceding numeral)	coin	max.	maximum
ca.	circa	min.	minimum
C.	century	mm.	millimeters
cm.	centimeters	N	north
D.	depth	NoEx (preceding numeral)	not from the excavations
diam.	diameter	P (preceding numeral)	pottery
dim.	dimension	P.diam.	preserved diameter
E	east	P.H.	preserved height
esp.	especially	P.L.	preserved length
est.	estimated	P.W.	preserved width
ext.	exterior	r.	right
G (preceding numeral)	glass	S	south
H.	height	S (preceding numeral)	sculpture
IN (preceding numeral)	inscription	T (preceding numeral)	terracotta
int.	interior	Th.	thickness
km.	kilometer	W	west
L (preceding numeral)	ceramic lamp	W.	width
L.	length	* (preceding numeral)	level (e.g. *98.00)
l.	left		
M (preceding numeral)	metal		

Dimensions of the glass objects are given in cm. throughout the volume.

Coin attributions that differ from those published in *BASOR* are on the authority of
Ann Johnston and T. V. Buttrey and will appear in the forthcoming monograph
on coins from Sardis.

Sector Abbreviations

For more complete sector information and explanation of Ac and AhT grids, see *Sardis* Rl (1975) 9-16; for site plan see Plan III.

Ac	Acropolis	BSH	S apsidal hall of central part of B
AcN	N spur of Ac		
AcS	S spur of Ac	BT	Bin Tepe Cemetery
AcT	Top of Ac	BW	Western area of B
AhT	Ahlatlı Tepecik	BWH	Central hall of BW
AT	Artemis Temple	B-W N Area	Northern part of BW
B	Building B, the Gymnasium complex	B-W S Area	Southern part of BW
		CG	Roman bath (formerly "City Gate")
BCH	Central hall of B		
BE	Eastern area of B	CG/MAE	Central arched recess (MAE) on the E side of the eastern wall (CGE) of Hall CGC in CG
BE-A, BE-B, BE-C, BE-D, BE-E	Rooms S of BE-H and BE-S		
BE-AA, BE-BB, BE-CC, BE-DD, BE-EE	Rooms N of BE-H and BE-N	CG/MAW	Central arched recess (MAW) on the W side of the eastern wall (CGE) of Hall CGC in CG
BE-H	Hall with pool W of MC		
BE-N	Room N of MC		
BE-S	Room S of MC	CW	City Wall
BK	Başlioğluköy	DU	Duman Tepe
Bldg. A	Unexcavated Roman building	EB	Eski Balıkhane
		ERd	East Road
Bldg. C	Roman basilica	Fc	see Syn Fc
Bldg. D	Unexcavated church, probably Justinianic	HoB	House of Bronzes and Lydian Trench area
BNH	N apsidal hall of central part of B	KG	Kâgirlik Tepe Cemetery
		L	Building complex SW of AT
BS E 1-E 19	Byzantine Shops, E shops numbers 1 through 19	LA	Altar of Artemis W of AT
		LNH 1-3	Long N hall N of Pa at B
BS W 1-W 16	Byzantine Shops, W shops numbers 1 through 16	M	Church M

MC	Marble Court at B (Gymnasium)	PHB	Hypocaust Building
MRd	Main Avenue (Main Road)	PIA	Pactolus Industrial Area
MMS	Monumental Mudbrick Structure	PN	Pactolus North area
		PN/E	Church E at PN
MTE	Middle Terrace East, trench S of HoB	PN/EA	Church EA at PN
		PT	Peacock Tomb
MTW	Middle Terrace West, trench S of HoB	R	Building R
		RT, RTE, RTW	Road Trench, Road Trench East, Road Trench West
NEW	Northeast Wadi		
NSB	Stelai bases set up along N side of LA	SSB	Stelai bases set up along S side of LA
Pa	Palaestra, E of MC	SWG	Southwest Gate
Pa-E, Pa-W, Pa-N, Pa-S	E, W, N, S corridors of Pa	Syn	Synagogue, S of Pa
PBr	Pactolus Bridge	Syn Fc	Forecourt of Syn
PC	Pactolus Cliff area	Syn MH	Main hall of Syn
PCA	Packed Columns Area E of Syn ca. E117–E124/S0–S4	Trench S	Trench S of AT
		TU	Acropolis Tunnels
		UT	Upper Terrace at HoB
PC LVC	Large vaulted chamber at PC	WRd	West Road
PC SVC	Small vaulted chamber at PC		

BIBLIOGRAPHY
AND
ABBREVIATIONS

Abbreviations of periodicals used throughout are those listed in the *American Journal of Archaeology* 82 (1978) 5–10.

The monographs and reports published by the Harvard–Cornell Expedition are cited under *Sardis*, below. The reports of the first Sardis expedition were published under the general series title of Sardis, Publications of the American Society for the Excavation of Sardis. Seventeen volumes were planned by H. C. Butler, Director of Excavations (*Sardis* I [1922] viii); of these, nine were actually published and are cited here under *Sardis*.

All publications preceding the first Sardis expedition will be found in the prospective *Bibliography of Sardis* (available from the Sardis Expedition, Fogg Art Museum, Harvard University, in mimeographed form). A preliminary selection appears in G. M. A. Hanfmann, *Letters from Sardis* (Cambridge, Mass. 1972) 346–349. Reports of the current expedition have appeared regularly since 1959 in the *Bulletin of the American Schools of Oriental Research* and *Türk Arkeoloji Dergisi* of the Turkish Department of Antiquities.

AJA 66   Axel von Saldern, "Glass from Sardis," *AJA* 66 (1962) 5–12.

Arbmann   Holger Arbmann, *Birka* I, *Die Gräber* (Stockholm [1943]).

Bagatti-Milik   Bellarmino Bagatti and J. T. Milik, *Gli scavi del "Dominus Flevit" (Monte Oliveto-Gerusalemme) Parte I. La necropolo del periodo Romano* (Jerusalem 1958).

Barag, "Glass Vessels"   Dan Barag, "Glass Vessels of the Roman and Byzantine Periods in Palestine" (Diss. in Hebrew, Hebrew University, Jerusalem, 1970).

Barag, "Netiv"   Dan Barag, "A Tomb Cave of the Byzantine Period near Netiv ha-Lamed He," *Atiqot*, Hebrew Series 7 (1974) 81–87.

Barag, *Shavei Zion*   Dan Barag, "The Glass" in M. W. Prausnitz, *Excavations at Shavei Zion: The Early Christian Church* (Rome 1967) 65–70.

Barkóczi   László Barkóczi, "Die datierten Glasfunde aus dem 3.-4. Jahrhundert von Brigetio," *Folia Archaeologica* 19 (1968) 59–86.

BASOR   *Bulletin of the American Schools of Oriental Research*

Baur-Kraeling   P. V. C. Baur, "Glassware" in Carl H. Kraeling ed., *Gerasa: City of the Decapolis* (New Haven 1938) 505–546.

Berger   Ludwig Berger, *Römische Gläser aus Vindonissa* (Basel 1960).

Brill   Robert H. Brill, "Glass Finds" in George M. A. Hanfmann, "The Fifth Campaign at Sardis (1962)," *BASOR* 170, 62–65.

Bruxelles Congrès   *Comptes Rendus II: VIIᵉ Congrès international du Verre, Bruxelles, 28 juin-3 juillet, 1965* (Brussels 1965).

Butler, *see Sardis*.

Calvi   M. C. Calvi, *I vetri romani del Museo di Aquileia* (Aquileia 1968).

Canivet   M. T. F. Canivet, "Vetri del V–VI secolo trovati nell'Apamene (Siria)," *JGS* 12 (1970) 64–66.

Clairmont   Christoph Clairmont, *The Glass Vessels:*

The Excavations at Dura Europos, Final Report IV, part V (New Haven 1963).

Crowfoot, *Samaria*   G. M. Crowfoot, "Glass" in J. W. Crowfoot et al., *The Objects from Samaria* (London 1957) 403–422.

Crowfoot-Harden   Grace M. Crowfoot and D. B. Harden, "Early Byzantine and Later Glass Lamps," *JEA* 17 (1931) 196–208.

Davidson, *Corinth* XII   Gladys R. Davidson, *Corinth* XII, *The Minor Objects* (Princeton 1952).

Delougaz-Haines   Pinhas Delougaz and Richard C. Haines, *A Byzantine Church at Khirbat al-Karak* (*OIP LXXXV*) (Chicago 1960) esp. 49, pls. 50, 59–60.

Edgar   M. C. C. Edgar, *Graeco-Egyptian Glass, Catalogue général des antiquités égyptiennes du Musée du Caire* (Cairo 1905).

Eggers   H. J. Eggers, *Der römische Import im Freien Germanien* (Hamburg 1951).

Eisen   Gustavus A. Eisen and Fahim Kouchakji, *Glass* (New York 1927).

Filarska   Barbara Filarska, *Szkła starożytne*, Muzeum Narodowe w Warszawie I–II (Warsaw 1952, 1962).

Fingerlein et al.   Gerhard Fingerlein, Jochen Garbsch, and Joachim Werner, "Die Ausgrabungen im langobardischen Kastell Ibligo-Invillino (Friaul). Vorbericht über die Kampagnen 1962, 1963, und 1965," *Germania* 46 (1968) 73–100.

FitzGerald   Gerald M. FitzGerald, *Beth Shan Excavations 1921–1923: The Arab and Byzantine Levels* (Philadelphia 1931) esp. 42–43.

Fossing   Paul Fossing, *Glass Vessels Before Glassblowing* (Copenhagen 1940).

Fremersdorf, *Fest. Oxé*   Fritz Fremersdorf, "Römische Gläser mit buntgefleckter Oberfläche," *Festschrift für August Oxé* (Darmstadt 1938) 116–121.

Fremersdorf, *Buntglas*   Fritz Fremersdorf, *Römisches Buntglas in Köln* (Cologne 1958).

Fremersdorf, *Naturfarbenes Glas*   Fritz Fremersdorf, *Das naturfarbene sog. blau-grüne Glas in Köln* (Cologne 1958).

Fremersdorf, *Gläser mit Fadenauflage*   Fritz Fremersdorf, *Römische Gläser mit Fadenauflage* (*Schlangenfadengläser und Verwandtes*) (Cologne 1959).

Fremersdorf, *Gläser mit Nuppen*   Fritz Fremersdorf, *Die römischen Gläser mit aufgelegten Nuppen in Köln* (Cologne 1962).

Fremersdorf, *Gläser mit Schliff*   Fritz Fremersdorf, *Die römischen Gläser mit Schliff, Bemalung und Goldauflagen aus Köln* (Cologne 1967).

Froehner, *Coll. Gréau*   Wilhelm Froehner, *Collection Julien Gréau: Verrerie antique . . . appartenant à M. John Pierpont Morgan* (Paris 1903).

Goldstein   Sidney M. Goldstein, "A Preliminary Study of the Glass Manufacture at Jalame in Israel," (Diss., Harvard University, 1970).

Haberey   Waldemar Haberey, "Spätantike Gläser aus Gräbern von Mayen," *BonnJbb* 147 (1942) 249–284.

Hackin   J. Hackin, *Nouvelles recherches archéologiques à Bégram* (Paris 1954).

Hamelin   Pierre Hamelin, "Matériaux pour servir à l'étude des verreries de Bégram," *Cahiers de Byrsa* (1953) 121–127.

Hanfmann, *JGS*   George M. A. Hanfmann, "A Preliminary Note on the Glass Found at Sardis in 1958," *JGS* 1 (1959) 50–54.

Hanfmann, "Découvertes"   George M. A. Hanfmann, "Découvertes archéologiques . . . Turquie," *Bulletin des "Journées Internationales du Verre"* 2 (1963) 124–126.

Hanfmann, *Letters*   George M. A. Hanfmann, *Letters from Sardis* (Cambridge, Mass. 1972).

Hanfmann-Saldern   George M. A. Hanfmann and Axel von Saldern, "Exploration of Sardis," *Institute of International Education News Bulletin* 36:9 (May 1961) 22–26.

Harden, *Karanis*   Donald B. Harden, *Roman Glass from Karanis Found by the University of Michigan Archaeological Expedition in Egypt, 1924–1929* (Ann Arbor 1936).

Harden, *Nessana*   Donald B. Harden, "Glass" in H. Dunscombe Colt ed., *Excavations at Nessana* (*Auja Hafir, Palestine*) I (London 1962) 76–91.

Harden, *ArchJ*   Donald B. Harden, "Ancient Glass, III: Post Roman," *ArchJ* 128 (1971) 78–117.

IEJ   *Israel Exploration Journal*

Isings   Clasina Isings, *Roman Glass from Dated Finds* (Groningen-Djakarta 1957).

Kisa, *Slg. vom Rath*   Anton Kisa, *Die antiken Gläser der Frau Maria vom Rath geb. Stein* (Cologne 1899).

Kisa, *Glas*   Anton Kisa, *Das Glas im Altertume*, 3 vols. (Leipzig 1908).

La Baume   Peter La Baume, *Glas der antiken Welt* Cologne, Römisch-Germanisches Museum. Wissenschaftliche Kataloge (Cologne [1973]).

Lamm, *Samarra*   C. J. Lamm, *Das Glas von Samarra* (Berlin 1928).

Lamm, *Mittelalterl. Gläser*   C. J. Lamm, *Die mittelalterlichen Gläser und Steinschnittarbeiten aus dem Nahen Osten* (Berlin 1929–30).

Lancel   Serge Lancel, *Verrerie antique de Tipasa* (Paris 1967).

Leciejewicz et al.   L. Leciejewicz, E. Tabaczynska, and S. Tabaczynski, "Gli scavi a Castelseprio nel 1962," *Rassegna Gallaratese di Storia e d'Arte* 24 (1965) 155–176.

Loeschcke, *Slg. Niessen*   Siegfried Loeschcke and H. Willers, *Beschreibung römischer Altertümer, gesammelt von Carl Anton Niessen* 3rd ed. (Cologne 1911).

London Congress   *Studies in Glass History and Design: Papers Read to Committee B Sessions of the VIIIth International Congress on Glass, Held in London 1st-6th July, 1968* (London 1968).

Morin-Jean   Morin-Jean, *La verrerie en Gaule sous l'Empire Romain* (Paris 1913).

Munich Congress   *Glastechnische Berichte* 32:VIII
(1959), *Sonderband: V. Internatl. Glaskongress* (Munich
1959).

Philippe   Joseph Philippe, *Le monde byzantin dans
l'histoire de la verrerie (V^e–XVIe siècle)* (Bologna 1970).

Puttrich-Reignard   Oswin Hans-Wolf Puttrich-
Reignard, *Die Glasfunde von Ktesiphon* (Kiel 1934).

Saldern, *Boston*   Axel von Saldern, *Ancient Glass in
the Museum of Fine Arts Boston* (Boston 1968).

Saldern, *Jb. Hamburg*   Axel von Saldern, "Sassani-
dische und islamische Gläser in Düsseldorf und Hamburg," *Jahrbuch der Hamburger Kunstsammlungen* 13
(1968) 33–62.

Saldern, *Slg. Hentrich*   Axel von Saldern, *Glassamm-
lung Hentrich: Antike und Islam,* Katalog des Kunst-
museums Düsseldorf I:3 (Düsseldorf 1974).

Saldern et al., *Slg. Oppenländer*   Axel von Saldern,
Birgit Nolte, Peter La Baume, and Thea E. Haevernick, *Gläser der Antike: Sammlung Erwin Oppenländer,*
exhibition catalogue, Museum für Kunst und Gewerbe
(Hamburg 1974).

Saller, *Bethany*   Sylvester J. Saller, O.F.M., *Excavations at Bethany (1949–1953)* (Jerusalem 1957).

Saller, *Moses*   Sylvester J. Saller, O.F.M., *The
Memorial of Moses on Mount Nebo* (Jerusalem 1941).

Sardis I (1922)   H. C. Butler, *Sardis* I, *The Excavations,* Part 1: *1910–1914* (Leyden 1922).

Sardis II (1925)   H. C. Butler, *Sardis* II, *Architecture,*
Part 1: *The Temple of Artemis* (text and atlas of plates,
Leyden 1925).

Sardis V (1924)   C. R. Morey, *Sardis* V, *Roman and
Christian Sculpture,* Part 1: *The Sarcophagus of Claudia
Antonia Sabina* (Princeton 1924).

Sardis VI.1 (1916)   E. Littmann, *Sardis* VI, *Lydian
Inscriptions,* Part 1 (Leyden 1916).

Sardis VI.2 (1924)   W. H. Buckler, *Sardis* VI,
Lydian Inscriptions, Part 2 (Leyden 1924).

Sardis VII (1932)   W. H. Buckler and D. M. Robin-
son, *Sardis* VII, *Greek and Latin Inscriptions,* Part 1
(Leyden 1932).

Sardis X (1926)   T. L. Shear, *Sardis* X, *Terra-cottas,*
Part 1: *Architectural Terra-cottas* (Cambridge, Eng. 1926).

Sardis XI (1916)   H. W. Bell, *Sardis* XI, *Coins,* Part 1:
1910–1914 (Leyden 1916).

Sardis XIII (1925)   C. D. Curtis, *Sardis* XIII, *Jewelry
and Gold Work,* Part 1: *1910–1914* (Rome 1925).

Sardis R1 (1975)   G. M. A. Hanfmann and J. C.
Waldbaum, *A Survey of Sardis and the Major Monuments
outside the City Walls* Archaeological Exploration of
Sardis Report 1 (Cambridge, Mass. 1975).

Sardis R2 (1978)   G. M. A. Hanfmann and N. H.
Ramage, *Sculpture from Sardis: The Finds through 1975*
Archaeological Exploration of Sardis Report 2 (Cam-
bridge, Mass. 1978).

Sardis M1 (1971)   G. E. Bates, *Byzantine Coins*
Archaeological Exploration of Sardis Monograph 1
(Cambridge, Mass. 1971).

Sardis M2 (1972)   J. G. Pedley, *Ancient Literary
Sources on Sardis* Archaeological Exploration of Sardis
Monograph 2 (Cambridge, Mass. 1972).

Sardis M3 (1975)   R. Gusmani, *Neue epichorische
Schriftzeugnisse aus Sardis (1958–1972)* Archaeological
Exploration of Sardis Monograph 3 (Cambridge, Mass.
1976).

Sardis M4 (1976)   C. Foss, *Byzantine and Turkish
Sardis* Archaeological Exploration of Sardis Monograph
4 (Cambridge, Mass. 1976).

Sardis M5 (1978)   A. Ramage, *Lydian Houses and
Architectural Terracottas* Archaeological Exploration of
Sardis Monograph 5 (Cambridge, Mass. 1978).

Simonett   Christoph Simonett, *Tessiner Gräberfelder*
(Basel 1941).

Smith Coll.   *Glass from the Ancient World: The Ray
Winfield Smith Collection,* exhibition catalogue, The
Corning Museum of Glass (Corning 1957).

Spartz   Edith Spartz, *Antike Gläser* Staatliche Kunstsammlungen Kassel (Kassel 1967).

Trowbridge   Mary L. Trowbridge, *Philosophical
Studies in Ancient Glass* University of Illinois Studies in
Language and Literature 13:3–4 (Urbana, Ill. 1930).

Vessberg, *Cyprus*   Olof Vessberg and Alfred West-
holm, *The Swedish Cyprus Expedition* IV:3 (Stockholm
1956).

Vessberg, *OpusArch*   Olof Vessberg, "Roman Glass in
Cyprus," *OpusArch* 7 (1952) 109–165.

Washington Congress   *Advances in Glass Technology* II:
*VIth International Congress on Glass . . . Washington,
July 8–14, 1962* (New York 1963).

Zahn, *Slg. Schiller*   Robert Zahn, *Sammlung Baurat
Schiller* Sales catalogue Rud. Lepke (Berlin, March
19–20, 1929).

ANCIENT AND
BYZANTINE GLASS
FROM SARDIS

INTRODUCTION

The glass published in this volume comprises over 900 items. With relatively few exceptions the finds of hollow glass (lamps, vessels etc.) consist only of fragments. Chronologically they fall into four periods.

Pre-Roman (**1–14**) includes a small series of sand-core vessels (in the following the terms "vessels" and "fragments" are interchangeable) and Hellenistic bowls. All recorded objects are included in the catalogue. *Roman* (**15–233**) consists predominantly of bottles and bowls from the first to the third or fourth century A.D. as well as a few other objects. Again all items identified as belonging to this group are included in the catalogue. *Early Byzantine* (**234–737**) dates from ca. A.D. 400 to 616. At least 95% of all glass finds from Sardis belongs to this period. Of those only a small portion—perhaps less than one tenth—has been listed because the publication of every sherd would have been redundant. We attempted to record those objects that are in fairly good condition—if one chooses to call the small but well preserved portion of a vessel an object "in good condition"—and that come from datable contexts. *Middle Byzantine* (**738–796**) consists of bracelets and "cakes" used as raw material, to the exclusion of almost any other objects. The period covered by these finds ranges from about the late tenth to the thirteenth and perhaps fourteenth centuries.

Two groups of objects (**797–874**) are not integrated within the chronological divisions. They encompass ring stones and beads, two series difficult to date when found unstratified and which have, therefore, been separated from the other objects.

By far the most important group of glasses from Sardis belongs to the Early Byzantine period.[1] The invaluable contribution of the Harvard-Cornell archaeological project was to furnish a large, relatively well dated body of material covering a period from about A.D. 400 to 616 which has enriched our knowledge of the history of post-Roman glass in the Near East.[2]

During the Imperial period numerous glass factories and small workshops had been established not only in the heartland of glassmaking—Syria and Egypt—and in Italy but also in practically all other regions under Roman rule. Thus manufacturing facilities were located along almost the entire border of the Mediterranean, in Asia Minor, South Russia, in the Roman-occupied territories north of the Alps, in Spain, and in England. These establishments produced all kinds of inexpensive table and kitchenware, profusely decorated luxury glass, window panes and mosaic cubes, beads and miscellaneous items such as game pieces, stamps, and rods.

Shortly before and after A.D. 400 when western Roman rule either diminished greatly or had ceased to exist altogether in the East as in the West, glass manufacture suffered a noticeable reduction in output and quality of work. Luxury glass was produced only in workshops within the Sassanian Empire to

---

1. Or late antique, cf. the terminology used in *Sardis* M4 (1976) ix.

2. For the dating of glass from Sardis weathered to form a layered structure in cross section cf. R. H. Brill, "The Record of Time in Weathered Glass," *Archaeology* 14 (1961) 18–22; R. H. Brill and H. P. Hood, "A New Method for Dating Ancient Glass," *Nature* 189:4758 (Jan. 7, 1961) 12–14.

serve the rulers and the wealthy; some of this glass appears to have been fashioned in the Euphrates region and was exported west. It includes extensively cut vessels of predominantly heavy clear or slightly greenish glass.[3] In most other regions in the Near East, particularly in Palestine, ordinary hollow glassware, windows, and tesserae were probably the sole products of workshops that catered to a clientele satisfied with simple glass utensils for everyday use. The earlier wealth of forms gave way to a more limited variety of shapes, decorative devices were kept to a minimum, and the total number of objects seems to have been much less than in the third and fourth centuries.

Factories within the boundaries of the Eastern Roman Empire continued to make glass that was very similar to that of the fourth century.[4] For example, the predominant fabric of the fourth century, the so-called Syrian blue green glass, continued to be used in the fifth and sixth centuries. Likewise basic forms such as long-necked bottles, wide-rimmed bowls, and stemmed goblets are found in various sites at levels datable to the late fourth as well as to the fifth and sixth centuries, without showing any appreciable change in style (see Chapter III, introduction).

The material found at Sardis is in accord with the general situation of post-Roman glassmaking in the East. The great importance of the finds, however, lies in the proof, documented by thousands of vessel fragments, that extensive manufacturing facilities in a provincial town such as Sardis probably began their operation shortly before A.D. 400, at a time when many factories in other regions throughout the ancient world had ceased production. Astonishingly large quantities of admittedly fairly ordinary hollow ware, flat glass, and tesserae were produced until the fall of the city in A.D. 616;[5] the finds thus provide us with an extensive body of material of a time that hitherto belonged to a "gray" area in the history of glass.

The Sequence of Glassmaking at Sardis

From Hellenistic times until the fourth century A.D. glass seems to have been relatively rare in Sardis. All the finds, even the ordinary bottles from graves, are probably imports. The cut and engraved ware of the luxury class satisfied the refined taste of the rich. Sometime in the very late fourth century at least one factory was established which supplied large amounts of table and kitchenware, flat glass for window panes, and tesserae. Extensive finds of cullet and fragments of pots are proof of such manufacture. These products seem to have mainly served the needs of the populace of Sardis. The glassmakers held fast to long-

established traditions developed particularly in Palestine; they do not seem to have been eager to experiment. And it appears that they were not asked by the citizens to produce vessels richly decorated with elaborate threading, engraving, cutting, molding or painting. Their products show such close affinities to glass made in other regions (Chapter III, introduction) that one is inclined to believe that Syrian glassmakers were responsible for at least the initial phase of manufacture in Sardis. In addition, relatively close contacts must have existed between the various establishments within the Byzantine Empire because the wares of the sixth and seventh centuries produced in diverse regions resemble one another to an astonishing degree. At Sardis, Early Byzantine glass comprises over three dozen basic forms, excluding their variants. Their publication may provide an important link between the better known late Roman and early Islamic glass. A few extraordinary finds such as the important fragment of dichroic glass (**63**) were certainly imports; Sardis does not appear to have had workshops equipped to turn out highly sophisticated glass vessels.

All activities must have ended in A.D. 616. Only much later, in the Middle Byzantine period, from the late tenth to the thirteenth or fourteenth century, one or more small shops again began to manufacture glass, but their products seem to have been limited to bracelets and maybe to window glass. A few stray finds of imported "Islamic" glass are of no consequence.

Terminology

Only a few words should be said about some of the terms used in this volume. Donald B. Harden (*Karanis*, 6–24) has set admirable standards for all writers following him by providing detailed terminology for ancient glass; the user of this volume should consult the introduction of the book on Karanis for a better understanding of the whole subject.

The lowest part of a vessel is the base. It can be flat or slightly concave (slightly domed), or it may have a pushed-up central part that can be conical in cross section ("high kick"). A base ring can be made by tooling the lowest part of the vessel in such a way that a double-walled ring is formed to serve as base. A sepa-

3. A. von Saldern, "Achaemenid and Sassanian Cut Glass," *Ars Orientalis* 5 (1963) 8ff.

4. See Barag, "Glass Vessels."

5. For other industries see *Sardis* M4 (1976) 14ff.

rately formed base ring can also be attached to the vessel.

Feet are usually made separately in conjunction with short stems and then attached to the bottom of the bowl. A foot as well as the rim of the bowl of a vessel may be left plain, or made more resistant by exposure to heat (fire polishing) whereby the rim receives a thickened, rounded edge. Edges of rims and feet can also be folded inward or outward; this operation sometimes causes an air trap within the folded edge.

Stems are cylindrical, concave (waisted), or conical (tapering upward or downward).

Vessel walls are either conical (straight-sided and tapering in one direction) or curved whereby the walls may, in profile, be concave or convex. Bottles are usually cylindrical, spherical, oval (with greatest diameter at center), ovoid (greatest diameter at shoulder), egg- or bell-shaped. If the exact shape of the body of a bottle or bowl is unknown the term" globular" is used.

A description of the colors and tints in ancient glass may also be found in Harden, *Karanis*. The colors of ordinary Roman Imperial glass show all shades of green, from a deep blue-green (aquamarine) to tints of green, olive-green, and yellow. As all pre-Roman and most probably also all Roman glass found in Sardis was imported, no clearly identifiable fabrics are distinguishable (ordinary bottles such as those cat-

alogued under **110–176** are practically identical to thousands of other bottles found around the Mediterranean that are greenish or yellowish in color). On the other hand, Early Byzantine glass, no doubt manufactured in Sardis proper, shows a remarkable homogeneity; it has, therefore, been possible to list a number of major fabrics (see Chapter III, introduction).

The glass found at Sardis shows many types of decomposition that range from a frosted, slightly iridescent surface to a scaling, many-layered scum and heavily pitted corrosion, the result of long exposure to adverse conditions in the soil. Many Early Byzantine pieces show a particular effect that is the result of intense heat during a conflagration: the glass surface is dark and dirty looking, covering a sometimes brilliant iridescence, an effect best described as "dirty fire scum." The most complete listing of all states of weathering and deterioration of ancient glass can be found in the introduction of Harden, *Karanis*.

Statistics

Listed in table 1 are only the vessels, fragmentary vessels, and fragments which have been included in the catalogue. In addition, the totals include all other vessel sherds that have actually been recognized as having belonged to identifiable forms. To the various totals was added a certain percentage (10–15%) to

Table 1. Pre-Roman, Roman, and Early Byzantine glass: estimated totals.

Period	Fragments		Vessels	Vessel totals	Windows
Pre-Roman Recorded and catalogued (actual no.)	28	=	28	28	
Roman Recorded and catalogued (est.)	119	and	66	185	
Early Byzantine Recorded and catalogued (inc. vessels)	420	=	420		Rough est.,
1958–60 (est.)	1000	=	500[a]		over 1000
1961–64 (est.)	4500	=	2300		panes
1965–66 (est.)	900	=	450		
1967–69 (est.)	500	=	250		
Total (est.)	7400	=	4000	4000	

[a] Cf. Chapter III, introduction. It can be assumed that no more than ca. 2 fragments represent one vessel, i.e. the number of fragments should be divided by 2.

take care of those fragments that were overlooked by the author. The glass fragments found in Sardis were recorded as follows: 1) Single objects which have an inventory number that have been entered in the registry. Most of these items are included in the catalogue. 2) Boxes containing more than one sherd coming from a specific location; sometimes half a dozen were put together in one box while in another case a box or carton might be filled with five or twenty-five pounds of sherds. All boxes are labeled with a locus. The author (a) selected out of these boxes single items for the catalogue, (b) counted those fragments in the boxes that he was able to identify and entered them in the statistical tables, (c) estimated the amount of nondescript fragments and also entered them in statistical tables (table 2). This task was complicated by the fact that much of the glass was smashed to small sherds and, therefore, resisted identification. It was equally difficult to estimate the total amount of unidentified sherds; however, it was felt that even very rough estimates, inaccurate as they may be, might give the reader a better understanding of the glass production in Sardis than no statistical table at all.

We should add that an attempt was made to record all pre-Roman, Roman, and Middle Byzantine glass. Because much of the material, particularly that from MTE and AcT, is unstratified, late Roman and Early Byzantine glass tends to get confused: common Roman Imperial ware of the fourth century is very similar indeed to common Early Byzantine ware of the centuries after ca. A.D. 400. Thus, many sherds of vessels and window panes computed in the statistical table as being Early Byzantine may actually be of third or fourth century date. This problem, however, is not as serious as one would think: as perhaps roughly 90% of the Early Byzantine glass is fairly ac-

Table 2. Early Byzantine types. The amount given for each type is based on sherds that were positively identified. These are minimum totals.

Objects	Number (est.)
Lamps	
Type 1	1[a]
Type 2	175
Type 3	140
Type 4	55
	400
Goblets	600
Ring bases	300
Concave bases (of bottles?)	250
Bottles, necks of bottles	250
Threaded necks	200[a]

[a] Cf. Chapter III, introduction. It can be assumed that no more than ca. 2 fragments represent one vessel.

curately dated, the margin of error with regard to confusing late Roman and Early Byzantine glass among the Sardis finds is very small. If a sure method of separating the two groups were to be devised in the near future, the totals as computed would not have to be changed by more than 1%.

All items with decoration, for example pattern molding and thread decoration, have been included in the catalogue. The majority of the sherds that could not be identified were probably part of bottles, bowls, lamps, goblets, beakers, etc., in this order. Thus, the total number of bottles as listed here (ca. 500) would increase considerably while beakers, on the other hand, may have existed in more limited numbers in Sardis.

I PRE-ROMAN GLASS

In first millennium B.C. levels at a site in Asia Minor one would expect a number of beads, a few vessels or fragments of core-formed glass, and perhaps the remains of pre-Hellenistic or Hellenistic bowls. The Sardis finds confirm what is known about glass excavated at other sites in Turkey.[1] Only very few finds of pre-Roman date were stratified. Thanks to recent archaeological work and extensive research much information has lately been gathered on core-formed and molded glass. While the former became old-fashioned toward the third and second centuries B.C., the latter, successor to the cut bowls of Achaemenid and early Hellenistic times, seems to have become the most popular glass vessel type during the second and first centuries.[2]

CORE-FORMED GLASS

Core-forming—the dominant manufacturing method in Egypt and Mesopotamia in the second half of the second millennium B.C.—reappeared in Mesopotamia in the eighth or seventh century B.C. and shortly afterwards along the Syrian coast.[3] The exact location of manufacturing facilities for the last group is as yet unknown; it can be assumed, however, that most of the workshops must have been situated in the area along the coast which was the heartland of Phoenicia and today comprises the western parts of Lebanon and Israel.[4] Other workshops appear to have been in operation on the Black Sea, in Rhodes, Cyprus, and elsewhere.[5] The frequency of production increased during the second half of the first mil-

lennium, to subside almost instantaneously when glass blowing was introduced. This event probably took place about 40–30 B.C. in precisely the same region that had witnessed the late flowering of core-

1. For example A. von Saldern, "Glass Finds at Gordion," *JGS* 1 (1959) esp. 34ff.

2. Among more recent surveys on Near Eastern glass of the millennium between 500 B.C. and A.D. 500, some with ample bibliographical references, cf. *Smith Coll.*, passim. D. B. Harden, "Ancient Glass, I: Pre-Roman," *ArchJ* 125 (1968) 46–72; idem, "Ancient Glass, II: Roman," *ArchJ* 126 (1969) 44–77. Saldern et al., *Slg. Oppenländer*, 8–16, 85–90.

3. For the most complete survey of core-formed glass see Fossing, passim. The core glass of the New Empire is treated in B. Nolte, *Die Glasgefässe im alten Ägypten* (Berlin 1968) while Mesopotamian and Syrian core glass prior to the mid-1st millennium was published by D. Barag in A. L. Oppenheim, R. Brill, D. Barag, and A. von Saldern, *Glass and Glassmaking in Ancient Mesopotamia* (Corning 1970) 129–199. The so-called late core-formed glass made in the Eastern Mediterranean from about the 6th–1st C. B.C. will be treated *in extenso* by B. Nolte and also by J. D. Cooney in two separate studies; for a recent survey on this group see B. Nolte in Saldern et al., *Slg. Oppenländer*, 13ff. For recent studies on the manufacturing methods used to produce core-formed glass see esp. M. Bimson and A. E. Werner, "Problems in Egyptian Core Glasses," *London Congress,* 121–122; idem, "Two Problems in Ancient Glass: Opacifiers and Egyptian Core Material," *Annales du 4ᵉ Congrès des "Journées Internationales du Verre," Ravenne-Venise, 1967* (Liège [1969]) 262–266; J. F. Wosinski and R. H. Brill, "A Petrographic Study of Egyptian and Other Core Vessels," *London Congress,* 123–124; D. Labino, "The Egyptian Sand-Core Techniques: A New Interpretation," *JGS* 8 (1966) 124–127. The term "core-formed" was introduced by D. Barag in" Mesopotamian Glass Vessels of the Second Millennium B.C.," *JGS* 4 (1962) 10.

4. Cf. D. B. Harden, *The Phoenicians* (New York 1963) 154–155.

5. B. Nolte in Saldern et al., *Slg. Oppenländer*, 13ff.

formed as well as molded glass.[6] Much of the production was exported; in fact the majority of finds come from Greece and the Greek islands, Spain, Italy, Asia Minor, etc.

A core-formed vessel was found in the 1912 campaign directed by H. C. Butler; nothing is known, however, about its shape or type of decoration as Butler's report lacks a description of the piece, which was lost in World War I.[7] Since 1961, five additional fragments of core-formed glass have been found. One, **1**, is fairly securely dated while the others come from disturbed levels. Because most are too small to allow a reconstruction of the shape of the original vessels, the core glass finds from Sardis will not add much information to the relatively uncertain chronological sequence of the types within the whole group.

Part of an alabastron, **1**,[8] was found (with an Ionian cup) in the tumulus, Duman Tepe, in the Lydian cemetery of Bin Tepe. It is datable to the first half of the fifth century B.C. **2** appears to have belonged to a fluted amphoriskos, a form common in the sixth and fifth centuries B.C.[9]

The most unusual fragment in the Sardis group, **3**, is made of opaque white glass with blue thread decoration, an extremely rare color combination. White core-formed glass, generally decorated with brown or purple thread patterns, is much rarer than blue glass with yellow, white, etc., threads and seems to be limited to the sixth and fifth centuries B.C. Much of the documented material in this series is known to come from South Russia.[10]

Most of the specimens in the core-formed group from Sardis have a yellow thread wound around the lip, a characteristic that does not help to date them more accurately nor to reconstruct their original form. Hellenistic core-formed glass differs from that of the mid-first millennium by its greater, or different, range of colors. A dark green material represented by **4** seems to be more in line with Hellenistic, than with sixth to fifth century glass.

1 *Pl. 1*. G66.8:7088. Ca. one-third of body of vertically fluted alabastron. Light turquoise-blue; yellow and white scale pattern.
H. 5.0.
BT/DU (inside chamber).
BASOR 186, 50–51, fig. 33.

2 *Pl. 1*. G61.25:3804. Curved section of core-formed, vertically fluted amphoriskos. Blue; decolorized, with 2 sections of zigzag threads in white and yellow.
Max. diam. 4.0.
PN S380/W250, level IV, ca. *88.00. 6th-5th C.; for dating of
BASOR 166, 24 (mention).

3 *Pl. 1*. G61.26:3882. Small, curved section (of alabastron or amphoriskos?). Opaque white; decolorized, blue zigzag pattern.
P.H. 1.9.
HoB E10/S90-S95, curved wall, to *98.40.

4 *Pl. 1*. Portion of out-turned rim of vessel. Black appearing dark green glass; applied yellow edge.
P.H. 1.0.
HoB 1963 E0/S115, well, to *92.12.

5 G66.9:7140. Flat fractured piece (of a core vessel?). Blue.
3 by 2, Th. 0.2.
BT/DU, chamber at mound 66.4, found with "straw embedded on one side."
BASOR 186, 52.

HELLENISTIC BOWLS

About ten fragments of Hellenistic glass bowls with lathe-cut rings were found. Fifteen years ago these vessels were thought to represent luxury glassware while the core-formed vessels, particularly at this stage in the development of glass, were certainly more ordinary. Recent finds, however, have shown that Hellenistic glass of the bowl type represented at Sardis was not as rare as once supposed. Derived from Achaemenid vessels, they are usually undecorated save for cut rings (cf. *Pl. 1*); those with embellishment show various kinds of cut leaf rosettes and, in rare cases, a shoulder frieze of "wings."[11]

6. According to most recent archaeological evidence the date of the introduction of the blowpipe may be advanced perhaps to the mid–1st C. B.C.; N. Avigad, "Excavations of the Jewish Quarter of the Old City of Jerusalem," *IEJ* 22 (1972) 200.

7. *Sardis* I (1922) 83; cf. Hanfmann, *JGS*, 53, "a complete specimen of figured glass commonly known as Phoenician."

8. Type: Fossing, 65–66, figs. 41–42, 5th C. Cf. Nolte in: Saldern et al., *Slg. Oppenländer*, 16 and nos. 184–185, here *Pl. 1*.

9. Ibid., 15; Fossing, 71–72.

10. Ibid., 70; Cf. Nolte, in Saldern et al., *Slg. Oppenländer*, nos. 153ff.

11. On Hellenistic glass, excluding core-formed glass, see G. D. Weinberg, "Glass Manufacture in Ancient Crete," *JGS* 1 (1959) 11–21; idem, "Hellenistic Glass Vessels from the Athenian Agora," *Hesperia* 30 (1961) 380–392; idem," The Glass Vessels," in *The Antikythera Shipwreck Reconsidered, TAPS* 55:3 (1965) 30–39; Saldern (supra n. 1) 34–49. Cf. also D. B. Harden, "The Canosa Group of Hellenistic Glasses in the British Museum," *JGS* 10 (1968) 21ff.; A. Oliver, Jr., "Millefiori Glass in Classical Antiquity," ibid., 48ff. Large quantities of plain Hellenistic bowls with cut grooves, dating from the 2nd and 1st C. were recently discovered at Tel Anafa (G. D. Weinberg, "Hellenistic Glass from Tel Anafa in Upper Galilee," *JGS* 12 [1970] 17–27), Kibbutz Hagoshrim (idem, "Notes on Glass

Within the group of undecorated bowls are three basic shapes with a number of variants: the hemisphere; a more shallow, segmental form; and a more pointed shape.[12] Their color is usually limited to light green—a few examples tend to be almost clear with green tint—and light brown (often called amber); in rare cases the glass is blue or purple. The plain bowls were probably made by enclosing ground glass between a negative and a positive mold and exposing this unit to a slow and extended heating process for fusing. After removal from the double mold the bowl was ground and polished.

Decorated bowls of this type always have one or more lathe-cut grooves located below the rim on the interior and, less frequently, on the exterior. Sometimes there is a cut ring at the interior bottom.

Bowls of this type have been excavated mainly along the Syrian-Palestinian coast and in Cyprus, Asia Minor, Greece, Italy, Egypt, and Southeast Europe. In addition, many examples without known provenance are preserved in public and private collections. Practically nothing is as yet known about the exact location of the manufacturing centers. We can assume that Alexandria, for one, must have been a center of luxury glass production where workshops specializing in cut, gold, and millefiori glass were situated.[13] However, other cities along the eastern half of the Mediterranean must also have harbored workshops making plain and cut glass bowls in the "Alexandrian manner." It is, therefore, conceivable that the examples found at Sardis were imported from any point along the Eastern Mediterranean.

Plain bowls seem to belong to the latter part of the Hellenistic period, i.e. towards the second and early first centuries B.C.[14] As far as I know no bowl has as yet been found that could be securely dated earlier or later than this period. Their manufacture was slowly discontinued when pillar-molded bowls (infra) came into fashion and glass blowing became the major manufacturing technique. Thus the latest example may have been made not later than the third quarter of the first century B.C.[15]

The fragments excavated at Sardis belong to the plain type. Most of them are brown, a few are green, and one, 7, has a purple cast. As the sherds are fairly small, no conclusion can be drawn as to the exact shape of the original vessels. However, there seems to be a preponderance of hemispherical and segmental shapes while the deeper, pointed type—generally perhaps the most frequent—may be represented only by one fragment which shows a slightly outsplayed rim, 6. Unfortunately, almost all fragments come from unstratified layers or from levels containing mixed Hellenistic and Roman material. Apparently no *terminus ante quem* could be established for any of the areas which contained remnants of Hellenistic bowls. With one single exception, 9, which was found in a pit in Syn MH with early Roman material (*BASOR* 182, 40), including part of a pillar-molded bowl, 27, the fragments of the bowls were excavated at HoB where they belong to relatively high levels.

6　G58.84:1010. Rim fragment of deep bowl; 2 grooves on int. below edge. Pale brown.
Max. dim. 3.5.
HoB room 9 ca. E12-E20/S53-S57, fill, level I, *98.50-98.30. Found with sections of similar bowls of greenish glass. For provenance see *BASOR* 154, 27–28.
AJA 66, 6, no. 2a, pl. 8:2a.

7　Rim fragment of hemispherical bowl; 2 cut int. grooves below edge. Pale purple, in sections turning to clear. Similar to 6.
Max. dim. 5.5, est. diam. 9.0.
HoB 1966 W5-W10/S105.

8　*Pl. 20.* G60.9:2714. Rim fragment of almost hemispherical bowl; cut int. groove below edge. Pale olive green.
Max. dim. ca. 5, est. diam. of bowl ca. 15 (very large).
HoB E10/S105 *101-100.5, unstratified.
AJA 66, 6, no. 2b, pl. 8:2b.

9　Rim fragment of segmental bowl; cut int. groove. Brown. Similar to 8.
Max. dim. 3.5.
Syn SE pit E87-E90/N1.5-N2.9 *94.40-94.

10　Rim fragment of segmental bowl; cut int. groove. Brown; heavily frosted.
Max. dim. 4.5, est. diam. ca. 13.
HoB 1961 E5-E10/S85 to *100.70.

11　Bowl fragment, similar to 10.
Max. dim. 4.5, est. diam. 15.0.
HoB 1963 W15/S110-S115 to *101.00.

from Upper Galilee," *JGS* 15 [1973] 36ff.), Ashdod (D. Barag, "Ashdod I," *Atiqot* English Series 7 [1964] 36–37), in Southeast Europe, and elsewhere.

12. Formerly believed to be Roman, the pointed form had mistakenly been thought to imitate the shape of a female breast, *Smith Coll.*, no. 237.

13. Cf. esp. Saldern (supra n. 1) 45ff.; Harden (supra n. 11); Oliver (supra n. 11).

14. Cf. particularly Weinberg, *Hesperia* (supra n. 11) 388–390.

15. Plain and ribbed bowls, the latter representing the first phase of pillar-molded bowls, were no doubt made side by side; cf. Weinberg, *JGS* 15, (supra n. 11) fig. 3.

12 Rim fragment of probably hemispherical bowl; cut int. groove. Brown; enamel-like scum.
Max. dim. 5.3.
HoB 1962 E5-10/S115 to *101.40.

13 G60.10:2715. Rim fragment of shallow bowl, cut int. groove below edge. Light brown.
Max. dim. 4.0.
HoB E20/S90, fill.
AJA 66, 6, no. 2c.

14 *Pls. 1, 20*. G60.11:2716. Two joining rim fragments of shallow bowl with curved rim. Broad cut ext. groove, ground edge. Clear with green tint.
Max. dim. 6, est. diam. 15, Th. 2.0.
HoB E15/S85, fill to *99.05.
AJA 66, 6, no. 2d, pl. 5:2.

II ROMAN GLASS

As is to be expected, the Roman glass from Sardis is more extensive than the pre-Roman. Graves and chance finds (**142–176**) have produced a few dozen quite ordinary bottles that are mainly datable to the first to third centuries A.D.[1] All shapes represented are well known from many Roman sites. With few exceptions the other Roman finds do not offer great surprises. As they all seem to be imports from Syria-Palestine, parallel finds abound—for example from Dura Europos. Remains of early Imperial colored ware, pillar-molded bowls, vessels with warts and thread decoration (including "nip't diamond waies") as well as common kitchenware were found in fair but not particularly unusual quantities. Material of this sort almost always includes a limited selection of stamps, pins, rods, rings, etc. (cf. the finds from Cyprus) and, occasionally, window glass.

The quantity of sherds of Roman vessels with elaborate decoration cut in geometric patterns that came to light was unusually large for a site in Asia Minor. They range in date from the late first to the third century and no doubt represent the luxury ware of the time. Exceptional is a rim fragment with an inlaid strip of blue glass, **60**. It seems that the wealthy clientele in Sardis purchased very elegant glass that suited their refined taste. Decorated glass—for example "zarte Rippenschalen," **42–44**—makes its appearance about the mid-first century A.D., and the latest cut glass is datable to the third century, suggesting that expensive objects imported from the southeast were in particular demand. In contrast, the period after A.D. 400, when glass was produced in large quantities in Sardis, seems to have been more austere: no luxury glass whatsoever is known to have been produced or imported!

EARLY IMPERIAL COLORED GLASS

An important category of early Roman Imperial glass comprises plates and bowls in an often strongly colored material typical of the luxury ware of the period. After fusing the glass in double molds, or after blowing, objects such as these vessels were extensively ground, generally on a lathe, to cut grooves into the surface and polish the exterior and interior walls. The shapes of the vessels are in most cases identical with, or very similar to, the contemporary terra sigillata ware. Most of the finds, especially those from Italy and Switzerland, can be dated to Tiberian and Claudian times. However, as the change from late Hellenistic to early Roman glass is almost imperceptible (cf. infra pillar-molded bowls), it is impossible to assign precise dates.

Most of the material in this group comes from Italy and southern Switzerland; however, finds like those from Sardis were certainly imported from regions along the Eastern Mediterranean. The objects are always well made and carefully finished, representing early Imperial luxury glass production and are com-

---

1. Cf. Isings, passim. For the change of the date of the invention of glass blowing to ca. 50–40 B.C. cf. N. Avigad, "Excavations at the Jewish Quarter of the Old City of Jerusalem, 1971," *IEJ* 22 (1972) 200.

parable to the contemporary mosaic and millefiori glass and the reticelli vessels.[2]

The small group of fragments of brown and blue bowls from Sardis belonging in this group is difficult to date. The sherds are too small to enable one to reconstruct the original vessels and none are dated by stratigraphic context.

Two pieces, **17** and **18**, come from the area where fragments of Hellenistic bowls were found. The other two, **15** and **16**, were discovered in Syn SE pit. A rim fragment of a purple bowl with grooves, **21**, found in a mixed fill, and a chance find of several fragments of a deep bowl in transparent turquoise-blue glass, **22**, appear to be closely related to the grooved vessels of colorless glass, **45–59**. Although the original vessels cannot be reconstructed from fragments of this size, it seems likely that they belonged to the first century A.D.[3]

Undoubtedly of early Roman date is the fragment of the bottom of a blue bowl with pronounced and ground base ring, **23**; it cannot be dated accurately with the aid of stratigraphy, but it is closely related to well-known types of the Julio-Claudian period.[4] With it was found the fragment of a dark blue cup, **24**, which can also be assigned to the first century A.D.; two other vessel sherds of deep aquamarine-colored glass discovered at the same spot are too small for identification.

Bowls

15 Fragment with 2 cut int. grooves, one wide, one narrow. Brown.
P.H. 3.0, Th. 0.5.
Syn SE pit 1965 E87-E90/N1.5-N2.5 *93.65.

16 Curved wall fragment. Brown.
P.H. 3.5, Th. 0.1.
Syn SE pit 1965 E87-E90/N1.5-N2.9 to *93.75.

17 Wall fragment, probably one cut groove. Brown.
P.H. 2.5, Th. 0.3.
HoB 1963 W20/S110-S115 to *100.80.

18 Wall fragment. Blue.
P.H. 3.5, Th. to 0.5.
HoB 1963 W5/S120-S125 to *101.20.

19 Group of vessel fragments of purple and brown glass. One rim fragment with int. groove, one with faint lines on ext. Found with 2 fragments of amber glass, belonging to late Hellenistic or early Imperial pillar-molded bowls.
HoB 1965, upper mixed fill, *102.40-100.50.

20 Curved wall fragment, one ext. cut groove. Light brown.
Max. dim. 3.8.
MTE 1964, ca. E65-E75/S150-S165 to *112.20. Found with **48**.

21 Slightly curved rim section with a wide and a narrow cut groove. Purple.
Max. dim. 4.1, est. diam. ca. 11.0.
HoB Lydian Trench 1963, upper mixed fill.

22 *Pls. 1, 20.* Several fragments, including 3 joining rim fragments probably of the same bowl. The rim fragments with slightly curved wall have a cut ridge 1 cm. below edge and a double groove 3 cm. below edge. Transparent turquoise-blue.
P.H. 6.5, est. diam. 10.0.
Chance find 1960.

Cups

23 Base section of cup, similar to **27**, with base ring and curving wall; concentric lathe marks on int. Blue.
Est. diam. base 4.2.
HoB 1964 W20-W25/S155-S160 *104.00. Found with **24**.

24 Flat base and lower portion of shallow cup or bowl with pronounced convex curvature. Dark blue.
P.H. 5.0.
HoB 1964 W20-W25/S155-S160 *104.00. Found with **23**.

PATELLA CUP

Profiled cups of colored glass always made with a base ring—so-called patella cups—derive from ceramic prototypes.[5] They are generally made of opaque white, blue, green, or red, as well as of millefiori, glass; cups of transparent glass are rare. Most of the cups known to us—many through the Eastern art market—appear to have been made in Syria; others

2. For a discussion of this type of glassware cf. particularly Berger, 24ff. D. D. Grose," Roman Glass of the First Century A.D., A Dated Deposit of Glassware from Cosa, Italy," *Annales du 6ᵉ Congrès de l'Association Internationale pour l'Histoire du Verre, Cologne, 1973* (Liège 1974) 31ff., 2nd quarter 1st C. Cf. also the mostly unpublished finds from Pompeii, *Smith Coll.,* nos. 173ff.; Froehner, *Coll. Gréau,* pls. 40ff.

3. For various grooved vessels of this period, for example, almost straight-walled beakers and bowls with slightly convex sides, cf. particularly Berger, no. 94ff.

4. Cf. ibid., 28, 30, no. 43, with parallel pieces cited.

5. Isings, no. 2. Cf. also *Smith Coll.,* no. 179 (no. 179f, here *Pl. 1*); Saldern et al., *Slg. Oppenländer,* nos. 278ff. Fremersdorf, *Buntglas,* pls. 29–30; La Baume, no. C 5.

were certainly manufactured in Italy and elsewhere in the late first century B.C. and first century A.D. (cf. *Pl. 1*).

The process of making them is similar to the method used for producing pillar-molded bowls (infra): ground glass was placed in a two-part, positive-negative mold and then fused; after cooling, the vessel was ground and polished.

The Sardis fragment, **25**, is made of clear glass, a fabric rarely used in this group. It was found in a level that included Roman ground and cut glass of the late first to third century; a facet-cut sherd, **64**, was discovered with this fragment.

25 *Pls. 1, 20*. Ca. one-half of body, curved wall with base ring, rim turned outward and upward. Clear; silver scum. P.H. 4.5; est. diam. rim 8.0, base 5.0.
MTE 1964 E67-E72/S157-S161 *111.90-111.70. Found with **64**.

SKYPHOS

A handle of aquamarine glass with horizontal thumb rest formed part of a skyphos, **26** (cf. also **209**). The earliest glass skyphoi—which imitated prototypes in metal and semiprecious stones—appear in late Hellenistic and early Roman Imperial contexts and belong to the luxury glass of this period.[6] Early Roman examples are still quite rare; they have either straight[7] or round walls (cf. *Pl. 2*).[8] Some of those made of blown glass and dated to the first century A.D. are executed in a less careful manner.[9] It seems likely that the handle from Sardis belongs to this latter group; it comes from an unstratified level (*BASOR* 162, 17).

26 *Pls. 2, 20*. G60.3:2393. Handle with thumb rest and portion of slightly curved wall. Aquamarine.
H. 2.7.
HoB E15/S115 *100.60, floor level.

PILLAR-MOLDED BOWLS

The so-called pillar-molded bowl is one of the most popular forms of early Roman Imperial glass. It is a segmental bowl of varying height with a frieze of more or less pronounced, vertically oriented ribs around the shoulder (cf. *Pl. 2*).[10]

Typologically the pillar-molded bowl succeeds the similar late Hellenistic bowl. The transitional stage between the two groups is represented by a number of bowls of roughly hemispherical shape and with rudimentary ribs that include, for example, a handsome piece (H. ca. 10 cm., diam. ca. 13 cm.) found in Masada (Jerusalem, Israel Museum); similar objects are found predominantly in Palestine.[11] The earliest examples of Roman Imperial pillar-molded bowls may have been introduced in the late first century B.C., probably in the Syrian-Palestinian region. They abound throughout the first century A.D. in practically all areas of the Roman Empire. One can distinguish a few variants which may have been made simultaneously; they include shallow and deep bowls, bowls with friezes of short and shallow ribs around the shoulder or with protruding and clearly defined ribs starting at the shoulder and tapering towards the base. It is as yet impossible to ascertain whether these variants follow a logical sequence in the development of form and decoration. However, one can be relatively certain that the majority of the finds belong to the first century A.D.

Pillar-molded bowls are generally made of greenish or pale aquamarine, more rarely of brown, blue, purple, and occasionally of opaque white glass; a number are made of millefiori glass.[12] The Sardis finds, no doubt imported from the Eastern Mediterranean, include seven fragments of aquamarine and pale green, four of brown, and four of blue, ratios not representa-

6. Isings, no. 39. Cf. also an early skyphos with later gold mounting, from Siverskaia: M. Rostovtzeff, *Social and Economic History of the Hellenistic World* (Oxford 1963) 600. For other skyphoi with carefully finished cut handles cf., for example, one recently found at Knidos, now Izmir Museum, and one in Corning Museum, *JGS* 13 (1971) 134–135, no. 3.

7. Isings, no. 39. For a recently published skyphos from Paredes see Maria A. Horta Pereira," O dolium cinerário, con skyphos vidrado a verde, da necrópole de Paredes (Alenquer)," *Conimbriga* 9 (1970) 49ff., pl. I. Cf. also J. and A. Alarcão, *Vidros romanos de Conimbriga* (Coimbra 1965) 46–48, nos. 55–56 (thumb rest).

8. A. Oliver, Jr., "Late Hellenistic Glass in the Metropolitan Museum," *JGS* 9 (1967) 30ff.

9. For example, Saldern et al., *Slg. Oppenländer*, no. 533; *Smith Coll.*, no. 165c.

10. Isings, no. 3. The literature on finds of pillar-molded bowls is extensive. The main types are listed by Isings. For typological predecessors cf. G. Eckholm, "De ribbade glasskålarnas ursprung," *Fornvännen* 1–2 (1958) 17–25. For a number of variants bought on the Eastern market cf. recently Saldern et al., *Slg. Oppenländer*, nos. 249–259. Cf. also Crowfoot, *Samaria*, 403ff. One method of manufacture (probably not the only one) is described by F. Schuler, "Ancient Glassmaking Techniques: The Molding Process," *Archaeology* 12 (1959) 47ff.

11. Cf. particularly G. D. Weinberg, "Notes on Glass from Upper Galilee," *JGS* 15 (1973) 35ff. Other examples are in the Museum Haaretz, Tel Aviv.

12. A number of millefiori bowls are listed in Saldern et al., *Slg. Oppenländer*, no. 328.

tive of the total number of finds of this type in either the East or the West.

The fragments found at Sardis belong to the most common variety, i.e. a shallow, segmental bowl with straight rim and pronounced ribs.[13] In general the shoulder zone along the termination of the ribs is carefully polished, one or two interior grooves are cut into the surface, and a cut ring often appears at the bottom. The manufacturing technique seems to have been identical or very similar to that of the Hellenistic bowls, **6–14**, namely melting ground glass between a positive and a negative mold; the ribs may have sometimes been applied after the basic vessel was completed.

The fragments listed here cannot be dated through stratigraphy. **27** and **37**, come from the same Syn SE pit that contained the fragment of a Hellenistic bowl, **9**. The majority were discovered in mixed levels at HoB which contained Hellenistic and Roman material.

27 Wall fragment with portions of 2 ribs. Pale aquamarine.
Max. dim. 3.7.
Syn MH SE pit 1965 E87-E90/N1.5-N2.9 *93.65.

28 Rim fragment, sections of 2 ribs; cut int. single groove and another pair of grooves 3 cm. below rim. Colorless with pale aquamarine tint.
Max. dim. 7.0, est. diam. 11.0.
Syn 1965 E87.5-E89/N3.25-N4.75 *92.60-91.60.

29 G60.12:2717. Rim fragment, small section of 2 ribs; pair of cut int. grooves. Aquamarine; slight decomposition.
Max. dim. 4.7, est. diam. 14.0.
HoB E15/S100, mixed Lydian to Roman fill.
AJA 66, 7, no. 4b.

30 Wall fragment, portions of 3 relatively flat ribs; pair of cut int. grooves. Colorless with bluish green tint and purple streaks.
Max. dim. 5.
HoB 1961 W10/S90-S95 to *101.15-100.35.

31 Wall fragment, sections of 4 ribs. Colorless with pale aquamarine tint; heavy brown scum.
Max. dim. 6.5.
HoB 1961 E5/S115 *101.80.

32 *Pl. 20*. Rim and shoulder fragment, several closely spaced ribs; 2 cut ext. grooves, one at int. Greenish.
Est. diam. at shoulder 15.0-17.0, Th. 0.5.
HoB 1963 W25/S105 to *100.20.

33 Wall fragment, sections of 3 ribs. Aquamarine.
Max. dim. 3.5.
HoB 1965 W34-W35/S120-S122 *102.00-101.40.

34 *Pl. 2*. G59.46:1530. Rim fragment, portions of 2 ribs, cut int. groove, rim heavily ground. Deep blue.
Max. dim. 5.0, est. diam. 18.0.
HoB near grave 59.g, E10/S60 *98.71.
AJA 66, 6-7, no. 4a, pl. 5:4.

35 Wall fragment, pair of cut int. grooves. Blue.
Max. dim. 3.5.
HoB 1963 ca. W13-W35/S120-S130, fill near surface.

36 Wall fragment, portion of rib. Blue.
Max. dim. 2.3.
PN 1965 W300-W305/S325-S328 *86.50-86.20.

37 Rim fragment, faint portion of rib, cut int. groove. Light blue.
Max. dim. 3.5.
Syn MH SE pit 1965 E87-E90/N1.5-N2.9 to *93.65.

38 Rim fragment, small portion of rib; pair of cut int. grooves. Light brown.
Max. dim. 3.5.
HoB 1963 W13-W18/S120-S125 to *102.30.

39 Rim fragment, sections of 2 ribs; pair of cut int. grooves. Light brown; heavy silver and brown scum.
Max. dim. 4.5.
MTE 1964 E60-E65/S149-S153 *110.00-109.50.

40 Wall fragment, sections of 2 ribs; pair of int. cut grooves. Light brown.
Max. dim. 3.0.
HoB 1968 W20-W22/S127-S132 *100.30.

41 Shoulder fragment, sections of 4 closely spaced ribs. Light brown.
Max. dim. 4.2.
HoB 1961 W5/S105, ramp to *99.90.

"ZARTE RIPPENSCHALEN"

The level at HoB containing fragments of Hellenistic bowls and early Roman material also included sections of three blown ribbed bowls of thin, pale purple, blue and light brown glass commonly referred to as "zarte Rippenschalen," delicately ribbed bowls of bulbous shape with short, outsplayed rim. Usually body and neck have a marvered-in, white spiral thread and pinched ribs. Opaque white with blue threads and

13. Isings, no. 3a.

clear examples are very rare. One group is plain, lacking the spiral thread in contrasting color.[14] A great number of the bowls were excavated in northern Italy but they also appear in Cyprus,[15] Syria, Asia Minor, and elsewhere; some are known through the art market.

The bowls were manufactured during a relatively short period of time; the majority of the dated finds belong to the second third of the first century A.D.

The two small fragments from Sardis, **42** and **43**, appear to belong to the undecorated group:[16] the sections preserved do not show any trace of white spiral threads. A fragment of the bottom of a blue bowl with white spiral threads, **44**, may also belong in this group. The level from which it comes can be associated with the early Imperial period (*BASOR* 174, 20–21).

42 *Pl. 20*. Shoulder fragment, with portion of vertical, highly raised and sharp, applied and pinched rib, with corresponding "bump" on int. Pale purple, almost colorless.
Max. dim. 4.0.
HoB 1963 W5/S115-S120 to *102.00.

43 *Pl. 2*. Shoulder fragment, portions of 2 ribs; out-curving rim. Light brown.
Max. dim. 4.0.
Chance find 1964.

44 Portion of concave base with end of spiral thread. Blue; white sunken threads, eggshell wall.
Max. dim. 5.0, diam. base ca. 4.5.
PN 1963 W228-W232/S349-S350 *88.40-87.90.

VESSELS WITH CUT AND ENGRAVED DECORATION

Vessels with Grooves

Fourteen recorded fragments were from vessels decorated with one or several grooves cut on the exterior wall. All the fragments are too small for exact identification but over half represent rim sections of either bowls or beakers and a number display profiles related to other well-known shapes, for example bottles.

This type of decoration most frequently occurs on bowls or cups of the first and early second centuries which are fairly deep and taper slightly towards the rim[17] and on conical beakers with straight walls, **45** and **47** (cf. *Pl. 2*).[18] Other vessel forms with cut

grooves do occur. Bottles with a globular body may have been among the original vessels at Sardis.[19] On the other hand, urns,[20] small bottles with tapering walls, short cylindrical neck, and conical rim made of clear[21] and ribbon[22] glass, do not seem to have been represented at the site. Numerous vessels with wheel-cut designs—facets for example—almost always show one or more grooves encircling the body; some of the Sardis fragments could be from such objects.[23]

The Sardis fragments are tinted in shades of green save for one fragment in blue, **50**, and two in purple glass, **54** and **55**.

Unfortunately, almost none of the Sardis finds are safely dated through archaeological means. The PN area where two of the vessel fragments were found near each other, **45** and **47**, includes the "Marble Monument" and a chamber tomb. The stratification of this area is complicated (*BASOR* 174, 22ff.). According to G. M. A. Hanfmann the floor level of the grave precinct whence the fragments come seems to be of the second or third century.[24] However, the tenuousness of such dating is clear when one learns that an Early Byzantine goblet stem was found near these fragments, **336**.

---

14. W. von Pfeffer and T. E. Haevernick, "'Zarte Rippenschalen'," *Saalburg Jahrbuch* 17 (1958) 76–91; T. E. Haevernick, "Die Verbreitung der 'zarten Rippenschalen'," *RGZM* 14 (1967) 153–166; Isings, no. 17. Cf. for recently published bowls of the undecorated and thread-decorated varieties Saldern et al., *Slg. Oppenländer*, nos. 260–266 (nos. 260, 262–263, here *Pl. 2*) and idem, *Slg. Hentrich*, nos. 35–37 (with parallels mentioned).

15. Cf., for example, A. H. S. Megaw, "Archaeology in Cyprus," *JHS-AR* 46 (1958) pl. 4f.

16. Cf. Saldern et al., *Slg. Oppenländer*, nos. 263ff.

17. Isings, no. 12, and Berger, nos. 94–95. Cf. Saldern et al., *Slg. Oppenländer*, no. 509, here *Pl. 2*. Such vessels predate the 4th C. conical beakers or lamps with grooves that occasionally bear applied prunts of colored glass (Isings, no. 106b:2); these beakers, in turn, either predate, or are contemporary with, faceted and grooved beakers of Sassanian origin (A. von Saldern, "Achaemenid and Sassanian Cut Glass," *Ars Orientalis* 5 [1963] 8, fig. 3). For bowl variants cf. also Davidson, *Corinth* XII nos. 588ff.; Berger, no. 101 and passim; Harden, *Karanis*, no. 330.

18. Isings, nos. 29–30 (for no. 29, see *Pl. 20*), 34; cf. also Davidson, *Corinth* XII no. 640.

19. Isings, no. 16, see *Pl. 20*.

20. *Smith Coll.*, no. 355.

21. Saldern, *Boston*, no. 26.

22. Isings, no. 7; A. Oliver, Jr., "Late Hellenistic Glass in the Metropolitan Museum," *JGS* 9 (1967) 23.

23. For example, faceted, conical beakers: Isings, no. 21; deep bowls with slightly curved walls: *Smith Coll.*, no. 354; bowls with rim constriction: Harden, *Karanis*, nos. 311ff.; Vessberg, *OpusArch* 114ff., pl. 1:18ff. Cf. also Calvi, 67, pl. C:9.

24. Cf. grooved beakers of 3rd C. date: Barkóczi, fig. 34:4, 5.

45 Rim and wall section of almost cylindrical vessel, 3 groups of 3 faint grooves each (2.5 cm. apart); ground rim. Clear, yellowish green tint.
P.H. 8.2, est. diam. ca. 8.0, Th. 0.25.
PN 1963 W225/S339-S342 *88.40-88.00. Found close to **47**.

46 *Pl. 20*. Curved wall fragment of beaker(?), ext. shallow cut groove flanked by engraved lines (two above, one below). Aquamarine; heavy iridescence, silver and brown scum.
Max. dim. 5.3, Th. 0.1.
MTE 1964 E65-E70/S155-S160 to *111.40.

47 Rim fragment of slightly conical vessel (beaker?), one groove 0.5 cm. below ground rim. Clear with green tint; flaking scum.
P.H. 5.5, est. diam. 10.0.
PN 1963 W225-W232/S340-S345 to *88.40. Found close to **45**.

48 Portion of cylindrical section of a vessel (beaker?). Two ext. cut grooves 1.5 cm. apart. Clear with green tint.
Max. dim. 4.3.
MTE 1964, ca. E65-E75/S150-S165 to *112.20. Found with **20**.

49 *Pl. 20*. Slightly inward curving rim fragment of deep bowl or beaker, faint cut grooves, below them a wide shallow groove. Aquamarine.
P.H. 7.0, est. diam. 12.0, Th. 0.15-0.20.
MTE/MTW 1964 (Roman upper debris), surface to *107.00.

50 Sprung rim fragment of bowl(?), wall curving inward at top, one broad ext. cut groove, an engraved line below. Blue.
Max. dim. 2.3, est. diam. 10.0, Th. 0.2.
MTE 1964 E65-E70/S155-S160 to *109.00-108.80.

51 *Pl. 20*. Sprung rim fragment of bowl or beaker, curving inward at top; one broad ext. cut groove, an engraved line below. Yellowish pale green.
Max. dim. 4.5, est. diam. 13.0, Th. 0.15.
MTE 1964 ca. E65-E75/S150-S165, S. part ca. *112.10.

52 Curved rim section of deep(?) bowl, 2 faintly cut lines at lower curvature. Aquamarine.
P.H. 4.3, Th. 0.3.
HoB 1963 W25-W35/S95-S115, upper mixed fill. Found with **54**, **55** and **67**.

53 *Pl. 2*. Slightly curved wall fragment of bowl or bottle, pairs of horizontal grooves, distance between each pair ca. 1.0. Almost clear.
Max. dim. 5.5.
UT 1959 E80/S170 ca. *118.70-118.20.

54 Slightly curved rim fragment of bowl(?), below rim a wide and a narrow groove. Purple.
Max. dim. 4.1, est. diam. ca. 11.0.
HoB 1963 W25-W35/S95-S115, upper mixed fill. Found with **52**, **55** and **67**.

55 As **54**, but with stepped groove below the rim.
P.H. 4.5.
HoB 1963 W25-W35/S95-S115, upper mixed fill. Found with **52**, **54** and **67**.

56 *Pls. 2, 20*. Rim fragment of hemispherical(?) bowl, edge flattened and ground, 3 wide ext. cut grooves. Greenish.
Max. dim. 4.5, est. diam. 11.0, Th. 0.3-0.4, W. grooves 0.5.
HoB 1962 W30/S95 to *100.40. Found with fragments of an Early Byzantine lamp (fabric 1) and a wide bowl (fabric 3).

57 *Pls. 2, 20*. Wall fragment of upper part of bottle or bowl with concave base, one ext. cut groove. Aquamarine.
Max. dim. 8.0, est. max. diam. of body below groove 10.0.
MTE 1964 E65-E70/S155-S160 *109.70.

58 *Pl. 20*. Wall fragment of upper shoulder of bottle, curves sharply inward at bottom, ext. groove. Clear with green tint.
Max. dim. 5.0, Th. 0.2, W. groove 0.7.
BS E 13, ca. E77.56-E81.96/S0-S5.
This fragment may either carry the wrong identification or may be a later intrusion. It is conceivable, however, that the sherd was part of a relatively late Roman Imperial vessel that had been preserved until the 5th C.

59 Wall fragment of bottle or bowl with curvature, pair of shallow lines. Aquamarine.
P.H. 6.0, Th. 0.15.
MTE 1964 E60-E65/S150-S155 *110.00-109.50.

Vessels with Grooves and Inlay

Apart from the faceted fragment of dichroic glass, **63**, the most exceptional fragment of Roman glass found in Sardis is the rim section of a beaker or bowl with inlaid decoration, **60**. A broad, cut groove below the edge serves as a channel for a set-in strip of translucent, turquoise-blue glass. The technical feat of putting this strip or band into the precut channel and slightly fusing the two surfaces is almost equal to the difficult overlay work achieved by the craftsmen responsible for cameo glass and the polychrome diatreta.[25] Decomposition has loosened the strip so that it can be removed.

25. Cf. for colored overlay glass, excluding the cameo glass, O. Doppelfeld, "Das neue Kölner Diatretglas," *Germania* 38 (1960) 403–417; G. Eckholm, "Orientalische Gläser in Skandinavien," *ActaA* 27 (1956) figs. 6:1, 10a; idem, "Glass Vessels with 'Cut Glass'

Another fragment found in the same year at UT, **61**, might conceivably come from a similar, or even the same, vessel. It has a channel of identical width below the edge, and the estimated diameter of the rim is very close to that of **60**. No trace of an inlay decoration could be discovered, however. No complete vessel with comparable decoration is known to me although there are numerous bowls and beakers with cut grooves encircling the region below the rim that could have received a strip of colored glass (cf. **45–59**).

The dating of these fragments is equally difficult. Both come from unstratified levels at UT. Most of the material discovered in this particular section is either late Hellenistic or early to mid-Roman Imperial (*BASOR* 157, 19–22). The profile of the rim sections, the way the grooves are cut, and the technique used suggest a date of the second to third century A.D.

60 *Pl. 2.* G59.33a(γ):1640. Rim section of bowl or beaker; below edge a broad groove or channel with an inlaid strip of pale turquoise-blue glass framed by 2 grooves. Because decomposition has loosened the contact between the clear and the blue glass, the blue strip can be taken off its bed. Very thin and clear, well made.
P.H. 4.2, est. diam. 10.0-12.0.
UT E80/S210 *121.55-119.90, unstratified.
AJA 66, 8, no. 7, pl. 5:7.

61 Rim section of upper part of bowl with straight wall, below rim broad cut groove (for inlay?). Almost clear, well made.
Max. dim. 5.0, est. diam. of rim ca. 12, Th. to 0.2.
UT 1959 E80/S210 *121.65-121.05. Found with **65**, **66**, **74** and **75**.

Vessels with Faceted and Geometric Decoration

No less than seventeen fragments of vessels with faceted and geometric decoration were found. This is an astonishingly large number for a site such as Sardis which is not rich in finds of luxury glass. The small size of the vessel portions recorded precludes, in most instances, a reconstruction of the original shape. None of the finds could be dated through their archaeological context but comparison with glass of the same type suggests that they belong to two periods, the early Roman Imperial period, and the third century.

Some come from areas at HoB that contained mixed Hellenistic and Roman material (*BASOR* 162, 12; 177, 15); others were recovered at UT which also is largely unstratified and included objects from Hellenistic through late Roman and Early Byzantine

times (*BASOR* 157, 20). Among the fragments, one appears to be unique because of its color, **63**; the remaining sherds can be associated with various groups of Roman cut glass of relatively high quality.

The first fragment, **62**, shows a faceted diaper pattern identical to that on a well-known group of straight-sided, conical beakers.[26] However, because this wall section is curved, it cannot have been part of such a beaker but certainly comes from a contemporary vessel type that has convex walls: a bowl of globular shape,[27] a bottle with spherical body,[28] a jug,[29] or a beaker of the Bégram variety with convex walls.[30]

Whether the next fragment, **63**, is late Roman or Early Byzantine—it was found in the forecourt of the

Patterns," *Viking* 20 (1956) 81ff.; J. Werner," Studien zu Grabfunden des V. Jahrhunderts" *Slovenská Arch* 7 (1959) 422ff. Cf. also A. von Saldern, "Glass Finds at Gordion," *JGS* 1 (1959) 44, no. 100.

26. Beakers were made mainly in a large (ca. 14 cm.) and a small (ca. 9 cm.) size; they are datable to the late 1st and early 2nd C.: Isings, no. 21; Eggers, nos. 185–187. Cf. esp. Hamelin, pls. 7–8; J. Werner, "Zu kaiserzeitlichen Glasbechern," *Germania* 31 (1953) 61–64; H. Norling-Christensen, "Hohe Glasbecher vom Pompeji-Typ mit einer Verzierung, die meistens aus eingeschliffenen dichtgestellten Furchen oder Facetten besteht," *Festschrift für Rudolf Laur-Belart, Provinciala* (1968) 410–427; cf. also Clairmont, 63ff. For other examples of this beaker type cf. also Harden, *Karanis*, 137ff.; Berger, 67ff.; Fremersdorf, *Gläser mit Schliff*, pls. 122–123; Saldern, *Boston*, no. 27; M. T. F. Canivet, *JGS* 11 (1969) 24, fig. 19. A beaker having a double groove interrupting the pattern at mid-body: *Guía de Museo y Necrópolis romana de Carmona (Sevilla)* (Madrid 1969) pl. 13. Mold-blown imitations of facet-cut beakers, dating from about the same time: Saldern et al., *Slg. Oppenländer*, no. 458 (add to the parallel examples cited: Tel Aviv, Museum Haaretz; Simonett, 84).

27. Facet-cut patterns arranged in a number of combinations and decorating a great variety of vessel forms are found in Fremersdorf, *Gläser mit Schliff*. The richest find of facet-cut vessels comes from Bégram, datable mainly to the 1st and 2nd C. They include conical beakers with grooves and ridges as well as with facets, bowls, jugs, beakers with convex walls, and a rhyton. For almost hemispherical wide-rimmed bowls cf. Hamelin, type C, pl. 7; a closely related piece in Tel Aviv, Museum Haaretz (*JGS* 7 [1965] 120, no. 3); Isings, 38; Saldern et al., *Slg. Oppenländer*, no. 511 (same shape but with widely spaced facets more typical of mid-Roman Imperial glass, here *Pl. 3*). For a more shallow variant with three rows of facets cf. Clairmont, no. 241.

28. Spherical bottles with short neck and wide rim: Fremersdorf, *Gläser mit Schliff*, pl. 118 (undated). For similar bottles with a combination of large circular cuts and smaller facets cf. Clairmont, no. 242, 61, n.148, listing parallel pieces mostly of the 3rd C. Spherical bottles with tube-shaped, constricted neck occasionally have facet patterns; they are mostly datable to the 3rd C.: Isings, no. 103; Saldern et al., *Slg. Oppenländer*, no. 514.

29. A series of wide, pear-shaped jugs come from Bégram: Hamelin, type B, pl. 7. For other jugs cf. Isings, no. 21, and Saldern et al., *Slg. Oppenländer*, no. 507; a jug of blue glass in Corning: *JGS* 8 (1966) frontispiece.

30. Hamelin, types D, F, pl. 8; Hackin, 254 no. 4, fig. 252.

Synagogue—cannot be decided at present. The considerable importance of this small wall section lies in the fact that it is made of a material of great rarity even in Roman times: it is dichroic glass, i.e. brownish green in reflected, and honey brown in transmitted, light. On its surface are portions of two groups of oval cuts above which are cut grooves. As in the case of the other fragments with curvature, the original object may have been a bowl, a beaker, or a jug.[31] While the facet pattern seems to be reminiscent of vessels of the late first and early second centuries bearing similar decoration (supra), the dichroic material relates the piece to the famous Lycurgus Cup in the British Museum, manufactured of almost identical glass.[32]

There seem to be only two other fragments of similar glass recorded, a fragment of a diatretum in the British Museum[33] and a facet-cut vessel fragment from Oxyrhynchus in the Victoria and Albert Museum.[34] Diatreta are generally thought to have been made in the fourth century, including the Lycurgus Cup and the fragment in the British Museum. The faceted fragment in the Victoria and Albert Museum, on the other hand, displays circles and curving bands with oval facets and is thought to be Islamic rather than Roman, i.e. apparently not earlier than the seventh century.[35] However, its cut pattern is unlike any designs found on Islamic cut glass at the height of its development, in the ninth to early eleventh centuries, but appears to be more in keeping with faceted and geometrically cut vessels of late Roman times.[36] If this proves correct, the Sardis fragment of dichroic glass should be dated in the fourth century, contemporary with the diatreta. Although its findspot apparently dates the piece within the Early Byzantine period, the Sardis fragment as well as the wall section from Oxyrhynchus seem to be products of workshops operating in the fourth century (in Egypt? Alexandria?) which were familiar with the difficult manufacture of dichroic glass and employed glass cutters able to produce diatreta.

Five other fragments bear similar facet patterns. Among them is only one rim fragment having a widely spaced facet design, **64**; its estimated diameter measures about 14 cm. Karanis finds of this type, also decorated with rows of vertically oriented facets, probably date from the late second or early third century.[37]

The other fragments with vertically oriented facets, **65** and **66**, may have been from vessels similar to the shallow bowl with curved rim, **64**. A more or less hemispherical bowl with facet pattern of a kind represented by the Sardis fragments was popular at Dura Europos in the second and early third centuries.[38]

The facet pattern on **67** is identical with that on a series of large vessel fragments from Dura said to have been part of beakers.[39] Some, or all, of these sherds may also have belonged to bowls with curved or slanted straight sides.[40] The unstratified Dura Europos fragments are dated to the early second century by C. W. Clairmont. By comparison with other finds, particularly those from Western Europe, the Dura fragments could also be of third century date.[41]

Also related closely to faceted vessels of the Dura Europos type is **68**.[42] A very small rim section, probably of a bowl, **69**, shows, below the rim constriction, the upper portion of a frieze of pointed arches with inscribed oval cuts. This motif occurs fairly frequently on a number of vessels which are predominantly third century, including bowls with a rim constriction similar to that of this fragment as well as that of **64**.[43]

The seven remaining fragments, **70–76**, belong to vessels with more varied cut decoration. They were either found at UT or in unstratified levels at HoB (cf. *BASOR* 162, 12–14). Of these seven fragments, four or five, **70–74**, appear to have come from vessels with very similar or even identical decoration. Their design consists of friezes of large circular or oval cuts framed by double lines; these motifs, in turn, are set within a framework of grooves and double grooves—

31. Cf., for example, Hamelin, pls. 7–8.

32. D. B. Harden, "The Rothschild Lycurgus Cup," *Archaeologia* 97 (1959) 179ff.; idem, "The Rothschild Lycurgus Cup: Addenda and Corrigenda," *JGS* 5 (1963) 9ff.

33. Harden, *Archaeologia* (supra n. 32) 188, no. B 7, pl. 69 a.

34. Ibid., pl. 69 b.

35. Ibid., 188.

36. Fremersdorf, *Gläser mit Schliff,* pl. 100.

37. Harden, *Karanis,* no. 211. Cf. also Fremersdorf, *Gläser mit Schliff,* pls. 74ff.

38. Clairmont, 63ff., figs. 2–3, esp. nos. 244 (allegedly from a beaker but probably from a bowl), 255 (with notes on parallel example). For variants of this type cf. also Harden, *Karanis,* no. 317 (segmented bowl with waisted lip); Fremersdorf, *Gläser mit Schliff,* pls. 75ff.; Saldern et al., *Slg. Oppenländer,* no. 511 (globular bowl with out-turned lip, here *Pl. 3*). For a skyphos with the same type of decoration cf. Fremersdorf, ibid., pl. 138.

39. Clairmont, nos. 244–247, pl. 25.

40. Fremersdorf, *Gläser mit Schliff,* 138, pls. 75ff.

41. Ibid.

42. Ibid.; Clairmont, 63ff., pl. 25.

43. Cf. the fragments of various vessels from Dura that show similar or identical motifs and are datable to the late 2nd and 3rd C.: Clairmont, 73ff., pl. 27; cf. also Harden, *Karanis,* no. 211, pl. 13. For a complete bowl of a type that may have been similar to the original Sardis vessel cf. one from Trier: Fremersdorf, *Gläser mit Schliff,* pl. 38. Variations of this motif decorate a number of vessels mainly of Western origin: Fremersdorf, *Gläser mit Schliff,* 55ff., 63, pls. 129ff., 192f.

apparently sometimes forming hexagons—which are accentuated by oval cuts. Among the closest parallels to these fragments are a series of finds from Dura Europos as well as a few other objects of Eastern provenance.[44] The original vessels to which the Sardis fragments belonged were undoubtedly bowls. However, the small size of the sherds does not allow a safe reconstruction of the bowl forms, which may have included both deep and shallow types. For example, the profile of the rim fragment, **72**, suggests a shape similar to a hemispherical bowl found in Cologne that is slightly waisted below the rim and displays a design of circular cuts accompanied by double grooves, gables, and small oval cuts. This piece as well as examples from Karanis are datable to the third century, although cutting of this type may have started in the second century.[45] Geometric cutting as described here appears to have reached its apex in the third century while diaper facet cutting (cf. **62**) was at its height in the late first and early second century.

The fragment of the bottom of a bowl, **75**, with a central, circular cut must have been part of a vessel having similar decoration. Many of the vessels with cut decoration found in the Near East as well as in the West show this or a similar motif at the base.[46]

Judging by the parallel material referred to above, the fragments with geometric, cut patterns should be datable to about the third, perhaps in some cases to the late second century. Unfortunately, none of the finds has been discovered in a stratified context. However, much of the material found with these sherds is of predominantly early or mid-Roman date consistent with the period suggested.

Although much Roman glass of good quality has turned up lately on the art market and is said to be from Asia Minor, it is highly unlikely that this material as well as the finds of luxury glass from controlled excavations was actually manufactured there. All glass objects of this type are imports from factories located on the Syrian-Palestinian coast or represent what Clairmont calls the East Syrian-Mesopotamian koine.[47]

62 *Pls. 3, 20.* G59.61:2122. Slightly curved wall section of cup, bowl, or bottle, narrow pattern of vertically oriented, oval facets. Clear with green tint; brilliant iridescence.
P.H. 3.8.
UT E80/S200 *122.85-122.00, unstratified.
AJA 66, 7, no. 5, pls. 5:5, 8:5.

63 *Pls. 3, 20.* G63.18:5785. Curved wall fragment of bowl, pattern of oval cuts arranged in rows. Pattern does not seem

to be continuous as an undecorated portion is visible on l. side; above are 2 lines and a 3rd ending abruptly (judged too shallow and, therefore, terminated?). Brownish green in reflected, honey amber in transmitted light (dichroic glass); pitting, scratches.
4 by 3, Th. 0.2-0.35.
Syn Fc E108-E110/N5-N9 *97.50-96.75, approximately the original floor level.
R. H. Brill, "The Chemistry of the Lycurgus Cup," *Bruxelles Congrès,* 223.1-13.

64 *Pls. 3, 20.* Rim fragment of bowl, convex wall with rim turned outward and inward (cf. **69**). On body, rows of oval cuts (one and portion of a 2nd row preserved); at rim, 2 cut lines, one line 1 cm. below edge. Clear, green tint.
P.H. 5.0, est. diam. 14.
MTE 1964 E67-E72/S157-S161 *111.90-111.70. Found with **25**.

65 Slightly curved wall fragment (of bowl or beaker?), narrow pattern of rows of vertically oriented, spike-like cuts. Almost clear.
Max. dim. 3.7, Th. 0.2, L. of facets 1.5.
UT 1959 E80/S210 *121.65-121.05. Found with **61, 66, 74** and **75**.

66 Slightly curved fragment (of bowl?), narrow pattern of rows of vertical, spike-like cuts. Almost clear.
Max. dim. 3.0, L. of facet 1.5.
UT 1959 E80/S210 *121.65-121.05. Found with **61, 65, 74** and **75**.

67 *Pl. 3.* Wall fragment (of bowl or skyphos?), cut facets. Preserved are portions of 4 rows: ovals, horizontal spikes, 2 rows of narrow ovals. Clear.
Max. dim. 5.1, Th. 0.3.
HoB 1963 W25-W35/S95-S115, upper mixed fill. Found with **52, 54** and **55** as well as with various small and highly deteriorated (unrecorded) fragments with remains of cut decoration and the section of a "zarte Rippenschale" (cf. **42–44**).

44. Clairmont, 70ff., nos. 265ff., n.182, datable to the 3rd C. A fragment with almost identical motifs was recently found at Pergamon (in storage at Pergamon). A hemispherical bowl with a deep groove below the rim (Smithsonian Institution, Washington, inv. no. 1929.8.170.43) is decorated with a very similar pattern. For pieces with similar patterns cf., for example, Froehner, *Coll. Gréau,* pls. 176:3; 178:4, 6; 181:3, 5. Cf. also Isings, no. 96b.

45. Cologne: Fremersdorf, *Gläser mit Schliff,* pl. 82, see also pl. 83 (here *Pl. 3*). Cf. the bowl in Washington (n.44 supra) and, for shape, the faceted bowl in Harden, *Karanis,* no. 317. The Cologne material is mainly of 3rd C. date; the Karanis find comes from one of the C-period houses, perhaps datable to the 2nd to mid–3rd C.

46. Cf., for example, Fremersdorf, *Gläser mit Schliff,* pls. 60, 101, passim.

47. Clairmont, 68, 71.

68 *Pl. 20.* Shoulder section of bowl, faceted surface. Preserved are portions of 2 rows of oval-hexagonal cuts, above them a groove and portions of 2 larger oval cuts. Clear, yellow-green tint; heavy, silvery iridescent scum.
P.H. 3.0.
MTE 1964 E65/S148 *108.60. Found with **69**.

69 *Pls. 3, 20.* Rim section of bowl, convex wall, rim turned outward and upward (cf. **64**). Below groove, frieze of pointed arches with inscribed ovals (carelessly executed). Clear with green(?) tint.
P.H. 3.5.
MTE 1964 E65/S148 *108.60. Found with **68**.

70 *Pls. 3, 20.* G59.44a:1810. Curved wall section of bowl. Preserved are portions of 2 sunken oval discs outlined by grooves and framed by a network of double grooves with oval cuts at intersections. Thick, clear, faint green tint; enamel-like scum.
5.3 by 4.0, Th. 0.45.
HoB E90/S110 *124.45-123.25, unstratified.
AJA 66, 7, no. 6b, pls. 5:6, 8:6b.

71 *Pl. 20.* G59.35a:1658. Two wall fragments of bowls (or a bowl) with curved walls. Preserved are portions of round and oval sunken cuts partly outlined by grooves and accompanied by slightly curved and straight cuts. Clear, faint green tint.
P.H. 4.8, Th. 0.4.
HoB E90/S110 *121.65-121.05, unstratified.
AJA 66, 7, no. 6a, pl. 8:6a.

72 *Pls. 3, 20.* Rim section of bowl. Preserved are portions of 2 sunken oval discs outlined by grooves, above is a groove with oval cut; below the rounded rim are 2 lines, one of which is positioned 1 cm. below edge. Clear, faint yellow tint.
P.H. 4.3.
HoB 1960 E0/S100-S100.5 *100.5-100.00. Found with **76**.

73 *Pl. 3.* Curved wall fragment of lower half of bowl (or bottle?). Pattern of large oval and circular cuts surrounded by a network of smaller cuts connected by grooves and double-grooves. Green tint.
Max. dim. 6.2, Th. to 0.4.
UT 1959 E80/S170, fill, *118.90-118.70.

74 Four non-joining fragments of straight-walled bowl with plain rim. Two grooves below rim; ca. 2.0 cm. below rim upper termination of cut decoration including oval facets and circular cuts. Almost clear, green tint.
Max. dim. of largest fragment 3.6, est. diam. of rim ca. 12.0.
UT 1959 E80/S210 *121.65-121.05. Found with **61**, **65**, **66** and **75**.

75 Portion of underside of bowl. Central circular cut surrounded by concentric grooves (2 preserved). Green tint.
Max. dim. 4.2.

UT 1959 E80/S210 *121.65-121.05. Found with **61**, **65**, **66** and **74**.

76 G60.51:3015. Rim section (of bowl?), pinched(?) honeycomb pattern, horizontal grooves. Pale purple.
Max. dim. 3.8, est. diam. 14.0.
HoB E0/S100-S100.5 *100.5-100.00. Found with **72**.

Vessels with Engraved and Scratched Decoration

Fragments of two glass vessels with crudely engraved or scratched decoration belong to this category, **77** and **78**. The manner in which the surface was scratched superficially with a pattern of shallow lines differs markedly from the carefully executed cut glass listed in the foregoing sections. Scratched decoration of this type, i.e. simple, carelessly drawn lines in a number of combinations, is mainly found on Roman glass of the third century. Cylindrical bottles with double handles and wide mouths appear to have been a favorite vehicle for this engraving.[48] Because the Sardis fragments show vertically oriented curvatures in their profiles, they cannot come from cylindrical bottles but must have belonged to bowls or pear-shaped or globular bottles the exact shape of which remains unknown.

The dating of the fragments is uncertain. **77**, comes from 1.65 m. below the *96.40 floor of BS E 15; **78**, was found at Pa-W, 1 to 2 m. below the level (*96.90) that included Byzantine glass (*99.80-97.70). In comparison with other glass having similar decoration the finds are datable to about the third or fourth century.

77 *Pl. 21.* G62.16:4744. Curved wall fragment (of bottle?), shallow engraved diagonal and vertical pairs of lines, carelessly executed. Greenish(?); dirty fire scum.
Max. dim. 5.0, Th. 0.25.
BS E 15 E89/S3 *94.75.

78 Curved wall fragment, irregular, scratched, horizontal lines. Green, bubbly; well-preserved.
Max. dim. 5.0, Th. 0.3.
Pa-W 1967 E41.30/N94.90 *96.90.

VESSELS WITH SQUARE ELEVATIONS

Among the most unusual glass finds of Roman date from Sardis are a bowl and a bowl fragment, **79** and

---

48. Fremersdorf, *Gläser mit Schliff*, pls. 32, 44, 59, 153ff.; Harden, *Karanis*, no. 431 (bowl); *Smith Coll.*, no. 375 (pear-shaped bottle with wide mouth); Saldern et al., *Slg. Oppenländer*, no. 516 (hemispherical bowl).

80, whose shape and quality of glass is very similar to the "zarte Rippenschalen," **42–44**. **79** is decorated with a band of carefully applied, almost square elevations around the shoulder. A slight "bump" on the interior of the bowl corresponds to each application on the exterior, an indication that the vessel was not blown in a mold.

This decoration brings to mind Hellenistic "winged" bowls and beakers[49] as well as a rare series of vessels with shields, ovals, rectangles, and kidney-shaped devices cut in high relief that are datable to the late first to early third century.[50] Only two parallel pieces are known to the author. Both are slightly lower and have a wider rim than the Sardis bowl. The first was found in Aquileia;[51] the other recently entered the Oppenländer collection.[52]

The more pronounced and larger elevations on the objects discussed here are related technically to the third century "vessels with warts," **81–90**, but also show affinities to even later Sassanian bowls which combine a row of pronounced, rounded elevations with pulled-out thorns or warts.[53]

As neither fragment from Sardis is stratified, a stylistic comparison remains the only way of assigning an approximate date. In view of their similarity to "zarte Rippenschalen," and their admittedly faint resemblance to some late Hellenistic vessels, we are inclined to attribute them to the late first or second century, preceding the vessels with warts.

79 *Pl. 3*. NoEx 71.19. Almost hemispherical bowl with concave base and out-turned rim, around shoulder a row of almost square elevations with corresponding slight "bumps" on the int. Clear with green tint; partly heavily weathered, broken and mended; sections, particularly at rim, missing. H. 7.2, diam. rim 7.5.

80 *Pl. 3*. G59.53:1620. Curved wall fragment of bowl, rectangular elevation at ext. with corresponding slight "bump" on int. Clear with green tint.
4 by 3.5; Th. 0.1, at elevation 0.7.
HoB E20/S210 *122.60-122.10, unstratified.
AJA 66, 6, pls. 5:3, 8:3.

VESSELS WITH WARTS

Types of Roman glass not yet systematically investigated include vessels with applied warts. Of the ten fragments found in Sardis that are decorated with applied and sometimes slightly pinched warts or droplets, only **81** is large enough to allow a tentative reconstruction of the original vessel. The sherd is part of a wall tapering upwards and turning at the bottom

sharply towards the center. Wall fragments of vessels with identical decoration have turned up at Dura Europos; they appear to comprise sections of bottles with inner diaphragms, and perhaps of bowls and beakers.[54] The fragment from Sardis may have been part of a pear-shaped bottle with short neck and probably a wide mouth; a base **187** of such a bottle or a similar vessel was found with **81**.[55] One can never be certain, however, whether such base sections belonged to bottles as described or to other vessel types such as beakers or bowls (cf. **185–187**). The original vessel of **81** could also have been either a relatively shallow[56] or spherical bowl with out-turned rim (cf. *Pl. 10*).[57] The latter form seems to have been particularly popular for this type of decoration. Such spherical bowls are approximately 6 to 8 cm. high, and their diameter varies from 8 to 12 cm. Some of the Sardis fragments, including **81**, could have belonged to vessels with these dimensions.

Bowls as well as bottles with applied warts, frequent in the Syrian-northern Mesopotamian region, are

49. Saldern (supra n.25) 38ff.

50. Ibid., n.100; idem, *Jb. Hamburg*, 33–36 (with notes on related pieces of early and mid-Roman Imperial date).

51. Calvi, 151, no. 320, pl. 24:3. The objects cited in n.57 are less closely related to the bowl than suggested by Miss Calvi.

52. Saldern et al., *Slg. Oppenländer*, no. 268. A deeper bowl with identical decoration: *The Constable-Maxwell Collection of Ancient Glass*, Sotheby's, June 4–5, 1978, no. 306.

53. Saldern, *Jb. Hamburg*, 46–48, fig. 7. Shinji Fukai, *Study of Iranian Art and Archaeology* (Tokyo 1968) figs. 30–33, color pl. 3, from Hassani-Mahaleh, tomb 7, undated.

54. Clairmont, 51ff. In n.105 Clairmont cites a bottle formerly in the Ray W. Smith collection (no. 322) as a parallel piece to the Dura fragments. In fact, however, the warts decorating the finds from Dura and Sardis differ markedly from the decoration on the Smith bottle although all objects in this category are closely related. The droplets on the Smith bottle are applied, pulled, and pinched while those on the sherds from Dura, Sardis, etc. are applied without being manipulated further. For mostly Western glass with warts cf. esp. Kisa, *Slg. vom Rath*, pl. 16:134–135, 137–139. Loeschcke, *Slg. Niessen*, nos. 195ff., pl. 23. Barkóczi, fig. 30:2 (spherical bottle, 3rd C.).

55. Cf. Zahn, *Slg. Schiller*, no. 283, pl. 9 (pear-shaped bottle, formerly von Gans collection: *Galerie Bachstitz* II, [The Hague 1921] no. 163, pl.67). For shape cf. *Smith Coll.*, nos. 312ff. (with snake-thread decoration); Eisen, 607, pl. 150.

56. Clairmont, no. 214.

57. For bulbous and hemispherical bowls with warts of Eastern as well as Western origin cf. esp. one example (at least) from Susa of late Parthian or early Sassanian date in Teheran, Archaeological Museum (storage; diam. ca. 12 cm.); a close parallel in London, Victoria and Albert Museum (*JGS* 11 [1969] 112 and *Burlington Magazine* [June 1971] 328). Cf. also Saldern, *Slg. Hentrich*, no. 107 (with bibliographic ref.); Kisa, *Slg. vom Rath*, no. 154, pl. 16:136; Loeschcke, *Slg. Niessen*, no. 216, pl. 27; Fremersdorf, *Gläser mit Nuppen*, pls. 34, 35 (here *Pl. 3*).

predominantly datable to the third century; the material from Dura Europos and Susa (n. 57, supra) may be assigned to the very late second and third century. Likewise, the Western finds have also been attributed to the third century. None of the Sardis finds comes from a dated context. They were discovered at UT and MTE in unstratified, mixed levels; in one case, **89**, the level contained a great variety of Roman and Early Byzantine material.

This decorative motif as well as the deep bowl form are also quite common in early Sassanian glass. In some instances it appears to be almost impossible to separate third century examples of Syrian provenance from Sassanian glass of slightly later date whereas the hemispherical bowls with slightly flaring rim, applied warts, or "bumps," and a register of pinched ribs no doubt belong to the following phase of Sassanian glassmaking, the fifth and sixth centuries.[58]

81 *Pls. 3, 21.* Wall fragment of cup tapering upward, 3 warts arranged in horizontal rows. Greenish, silver iridescence.
P.H. 5.5, est. max. diam. 10.0, Th. to 0.3.
UT 1959 E80/S200 *121.50-121.00. Found with **187**.

82 Curved section of bowl, applied warts. Green tint.
Max. dim. 4.0.
HoB 1959 E80/S20 *123.50-123.00.

83 See **82**.
P.H. 3.0.
HoB 1961 E0-W5/S110 to *102.10.

84 See **82**.
P.H. 4.0.
HoB 1962 E5-E10/S115 to *100.40.

85 See **82**.
P.H. 4.0.
MTE 1964 E65-E67/S153-S154 *111.00-110.00. Found with a few small vessel fragments of about the same period: clear with green tint and heavy silver iridescence; among them are a bottle neck flaring upward and a rim (of a flat dish?) profiled to form a slight elevation on top.

86 G64.1:6044. See **82**.
P.H. 4.0.
MTE E60-E65/S150-S153 to *109.90.

87 See **82**.
P.H. 4.0.
MTE 1964 E55-E60/S143-S145. Found with the base of a bottle showing a mold-blown rosette, probably an Islamic or even later intrusion, and Early Byzantine vessel fragments.

88 See **82**.
P.H. 3.0.
MTE 1964 E60-E65/S145-S150 to *110.00. Found with various fragments of Roman and Early Byzantine date.

89 See **82**.
P.H. 2.5.
HoB 1965, upper mixed fill, *102.40-100.50. Found with many glass fragments, particularly of the Early Byzantine period, including ring bases (cf. **401-444**), portions of goblets, bottles, and lamps.

90 See **82**.
P.H. 2.0.
HoB 1965 W34-W35/S120-S122 *102.00-101.40.

VESSEL WITH PINCHED DECORATION

Only one globular bottle with a row of pinched warts was found, **91**. This type of decoration, related to the applied warts, appears mostly on bottles and bowls made predominantly in the Syrian-Mesopotamian region in the third and fourth centuries.[59]

91 *Pl. 4.* Fragmentary, almost spherical bottle, concave base, short neck, and funnel mouth with infolded rim; band of pinched nipples around shoulder. Pale aquamarine, eggshell.
H. 12.0, diam. rim 4.0.
Özbek Tepe 1964, chance find.

VESSELS WITH "NIP'T DIAMOND WAIES" DECORATION

Two vessel fragments show a decoration of horizontally oriented double-threads pinched together at regular intervals to form a chain-like design (cf. *Pl. 4*). In seventeenth century English glass this sort of pattern is commonly referred to as "nip't diamond waies."

Such decorative devices are fairly common in Syr-

58. Cf. Saldern, *Jb. Hamburg,* 46-48. For other examples found in Iran (undated): Fukai (supra n.53) pls. 30ff. Various examples found in Iraq in Baghdad, Iraq Museum (inv. nos. IM 22364, 33666, 33667). For vessels with applied warts of 11th–12th C. date cf. Davidson, *Corinth* XII 83ff. and G. D. Weinberg, "A Medieval Mystery: Byzantine Glass Production," *JGS* 17 (1975), 136ff.; D. B. Harden, "Some Glass Fragments, Mainly of the 12th–13th Century A.D., From Northern Apulia," *JGS* 8 (1966) 70ff.
59. For a bottle similar to the one from Sardis: Spartz, no. 149 (with bibliographic ref.). Cf. also Saldern, *Slg. Hentrich,* nos. 114ff; Eisen, fig. 259.

ian glass of the third and fourth centuries. The closest parallels to the Sardis fragments are a number of sherds from Dura Europos.[60] The first of the Sardis finds, **92**, is bent sharply outward at the top which may be an indication that it was part of a deep(?) bowl or beaker with the rim turned outward.[61] It was found in a pit in BE-A which contained coins, the latest of which are safely identifiable as those of Constantius II (A.D. 355–361; *BASOR* 187, 12–13). The other fragment, **93**, comes from an unstratified context.

Most of the related material with applied and pinched ribbing is datable to the third century. As **92** must be at least earlier than the mid-fourth century, it is quite likely of third or early fourth century date.

92 *Pls. 4, 21.* Wall fragment, turning sharply outward at top (or bottom?), portion of applied and pinched thread pattern. Clear, green tint.
Max. dim. 4.0.
BE-A 1966 E11.90-E13.90/N7.90-N9.90 pit *96.80-96.30.

93 Three fragments (of one vessel?), one plain, two with portions of "nip't diamond waies." Aquamarine.
P.H. (largest fr.) 4.0.
MTE 1964.

VESSELS WITH CRIMPED BANDS APPLIED TO THE RIM

A series of rim sections of vessels with crimped, wavy bands attached along the edge have come from various locations at Sardis. Almost all of them are of aquamarine glass. The rims are folded, containing air traps; the crimped bands or edges are formed in an identical fashion which seems to be an indication that they all come from similar or identical vessels. The estimated diameter of the rim of such an object is about 10 to 12 cm.; the preserved lengths of the sections with crimped bands measure 3 to 6 cm. Since the whole group is fairly homogeneous, all sherds recorded are likely to date from about the same period.

Crimped bands such as these generally belong to shallow or deep bowls (*Pl. 4*). The bands are attached in pairs at opposite sides of the rim. The bowl form derives from terra sigillata ware and appears to have been common in glass from the late first to the third century in Syria, Cyprus, and elsewhere.[62]

Two of the crimped sections at Sardis were found in contexts that might allow approximate dating of the objects. One, **94**, comes from an area at PN (east side of unit N) that contained Roman burials and had been frequently rebuilt until Early Byzantine times

(*BASOR* 174, 22–24; *Sardis* M4 [1976] 46-48). The other, **95**, comes from a pit dug at the southeast corner of the main hall of the Synagogue. At a depth of *93.80—the rim was found at *93.75—there are walls of a late Hellenistic or very early Roman Imperial, no doubt constructed before the earthquake of A.D. 17 (*BASOR* 182, 40). As the original vessel certainly must have been made after this date, the fragment can only represent an intrusion.

The other rim sections were part of unstratified levels at HoB. One, **96**, was found in a mixed fill in 1960 (*BASOR* 162, 12), and the others, **97–100**, were found in the "enormous dump" with material ranging from late Hellenistic to late Roman times (*BASOR* 177, 15).

The average length of **94–101** is 3.0–6.0; est. diameter of vessels 10.0–20.0.

94 *Pl. 21.* Green tint.
PN 1963 W220/S350-S352 to *89.50.

95 Aquamarine.
Syn MH SE pit 1965 E90-E91/N1.5-N2 *93.75.

96 *Pl. 4.* G60.14:2719. Aquamarine.
HoB E10/S100 *101.50-100.00.

97 *Pl. 4.* G64.4a:6157. Aquamarine.
MTE E62-E65/S146 *108.50.

98 Greenish.
MTE 1964 E65-E70/S155-S160 *110.90.

99 Aquamarine.
MTE 1964 E60-E65/S149-S153 *110.00-109.50.

---

60. Clairmont, nos. 194ff. (group 2) which are datable, together with related material (nos. 186ff.), to the 3rd C. For various vessels of Eastern provenance with this type of decoration cf. Eisen, fig. 160, pl. 112; *Smith Coll.,* no. 305; Barag, "Glass Vessels," pl. 13; Harden, *Karanis,* no. 593. For Western examples cf. Morin-Jean, 51, n.1, figs. 264–266; Fremersdorf, *Gläser mit Fadenauflage,* pls. 108f., 114ff. (3rd and early 4th C.). Cf. also D. B. Harden, "Four Roman Glasses," in: Joan Liversidge, "Roman Discoveries from Hauxton," *Proceedings of the Cambridge Antiquarian Society* 51 (1958) 7–17.

61. Perhaps similar to a footed beaker with "nip't diamond waies" (*Smith Coll.,* no. 305) or to cut bowls (Harden, *Karanis,* nos. 212ff.), all belonging to the 3rd C.

62. Isings, no. 43, here *Pl. 21.* Vessberg, *OpusArch,* 112–113, pl. 1:16; idem, *Cyprus,* figs. 42:16, 43:6. For 2nd and 3rd C. rims of this type found in Crete cf. T. S. Buechner, "The Glass from Tarrha," *Hesperia* 29 (1960) 112–113, no. 16. Cf. also J. Alarcão, "Vidros romanos de Museus do Alentejo e Algarve," *Conimbriga* 7 (1968) nos. 33–34.

100 Two fragments: a) aquamarine; b) greenish(?); black scum.
MTE 1964 E60-E65/S145-S150 to *107.20.

101 Aquamarine(?); dark scum.
AcT trench E 1960 E-F/20-27, fill *397.00.

APPLICATION

One disc-like application, **102**, from a vessel wall was found in Sardis; its exact findspot is unknown. It may come from a bottle belonging to a group of Syrian third century glass vessels with applied, disc-like motifs.[63]

102 Circular stamp-like application(?) with portion of curved vessel wall, stamp tooled to form 3 parallel grooves. Aquamarine.
P.H. 3.5; diam. stamp 2.2, D. 0.5; Th. vessel wall 0.1.
Chance find 1962.

VESSELS WITH APPLIED SPIRAL THREAD DECORATION

Sections of two early Imperial vessels with applied spiral thread decoration came to light. One, **103**, was part of a small object, probably an amphoriskos.[64] The level in which it was found suggests an early Roman Imperial date.

The other vessel fragment, **104**, appears to have been part of a cup belonging to a well-known group of the first century, the many examples of which are usually left undecorated or bear faint engraved lines.[65]

103 G61.27:3910. Section of short neck and shoulder of small, blown amphoriskos(?). Decolorized; applied yellow spiral threads.
P.H. 2.6.
PN W250/S370 *88.58.

104 *Pls. 4, 21.* G60.36:2761. Quarter of rim and slightly, upward-tapering wall of a cup, 2 applied threads. Aquamarine.
P.H. 4.8, est. diam. rim 14.0, Th. 0.1-0.15.
AcN E0-E16/S1-N3, high fill, ca. *394.00.

BEAKERS

A slightly bell-shaped beaker, **105**, with a thread applied below the rim and a clearly defined, double-walled foot was found in grave 67.18 at Ahlatlı Tepecik (*BASOR* 191, 9–10). While the eggshell glass and

the thread around the rim indicate a date in the first century,[66] its relatively high, slender shape is more typical of the second century.[67] The coins found in this series of graves are of the first to second century; the beaker probably dates to the latter part of this span.

Part of the upper portion of a beaker or bowl, **106**, may belong in this series. Its straight sides seem to indicate a slightly conical or almost cylindrical shape. The outer surface bears faintly engraved lines which run in a diagonal direction (the lines do *not* appear to be marks of applied threads broken off). This fragment is similar in shape to the upper portions of those beakers which occasionally show faint, often roughly engraved lines arranged horizontally.[68] Because the fragment from Sardis comes from unstratified MTE, its date cannot easily be ascertained. However, we can tentatively assign it to the third or fourth century.

Two other bases, **107–108**, may belong to beakers. Base **107** might have belonged to a beaker similar to **105** since the construction of its foot and lower body are the same.

The profile of **108** is close to the bases of beakers of the first or second century[69] but could also be compared with bases of bowls of the second and third centuries.[70]

---

63. Cf. *Smith Coll.,* no. 326. For vessels with applications bearing a stamped image of a mask, a lion's head etc. cf. for example, *Smith Coll.,* no. 192 (jug); R. Joffroy, "Note sur une oénochoé trouvée à Vertault (Côte-D'Or)," *Bulletin de la Société Archéologique et Historique du Châtillonais* 3d ser. no. 10 (1958), 282–284. For broken-off, stamped applications cf. *Smith Coll.,* no. 193; Loeschcke, *Slg. Niessen,* no. 162, pl. 39; Filarska, pls. 17ff.; Saldern et al., *Slg. Oppenländer,* nos. 529ff.

64. Perhaps of a type similar to *Smith Coll.,* no. 356 (with engraved lines). A bottle with a profile similar to the Sardis fragment, decorated with spiral thread: Vessberg, *OpusArch,* 133–134, pl. 7:19; cf. also Saldern et al., *Slg. Oppenländer,* no. 642.

65. Isings, no. 12. A few of the cups have enameled or painted decoration: Saldern, *Slg. Oppenländer,* no. 397 (with bibliographic ref.).

66. For its shape cf. Isings, no. 35, citing examples from Pompeii.

67. Stylistically the Sardis beaker should be placed between 1st C. pieces (Eisen, pl. 65 center, from Tripoli; Zahn, *Slg. Schiller,* no. 221, pl. 12; Barag, "Glass Vessels," pls. 19, 32) and those datable to the 2nd–3rd C. (Vessberg, *OpusArch,* 119, group "B II," pl. 4:6, 8). At least one beaker-type of the late Roman Imperial period tends to have straight walls: La Baume, nos. D 100, D 101, pl. 40. For a group of beakers of various forms cf. Loeschcke, *Slg. Niessen,* pl. 44.

68. For example, La Baume, no. D 101, pl. 40. Cf. also Isings, no. 106, type c:2.

69. Vessberg, *OpusArch,* pl. 4:11; Barag, "Glass Vessels," pl. 19, bottom.

70. Vessberg, *Cyprus,* 170, fig. 50:40, 41.

105 *Pls. 4, 21.* Slightly conical beaker with sloping folded foot, applied thread below plain rim. Aquamarine, eggshell; various sections of body missing.
H. 14.5, diam. foot 5.2, rim 8.3.
AhT 1967 grave 67.18, W9.95-W10.05/S50.75-S50.85 *74.95-74.92.

106 *Pl. 21.* Portion of upper part of beaker (or bowl?), cylindrical, faintly visible and roughly executed engraved lines in diagonal pattern. Clear with green tint, eggshell.
P.H. 4.2, est. diam. rim 13.0.
MTE 1964 E64-E72/S155-S161 to *112.20.

107 *Pls. 4, 21.* Lower portion of beaker(?), pad base fused to vessel with slightly convex wall. Aquamarine.
P.H. 2.8, diam. base 4.1.
HoB 1963 W5-W10/S120-S125, fill, *101.80.

108 *Pls. 4, 21.* Lower portion of beaker(?), base ring with air trap, wall faintly tapering downward. Green tint, eggshell.
P.H. 3.8, diam. base 6.0.
HoB 1963 W5-W10/S120-S125, fill, *101.80.

109 *Pl. 21.* Slightly concave base of beaker or cup; perpendicular bottom part beginning to increase in diam. at fracture. Aquamarine.
Diam. 4.5.
HoB 1963 W13-W25/S120-S130 to *102.00.

BOTTLES WITHOUT DECORATION

Bottles are among the most common glass vessels of Roman times. The finds from Sardis are mainly the simplest types; among them are many vessels commonly referred to as "tear bottles" or unguentaria. Closely related to these finds is a group of bottles acquired from a villager in 1963, **142-176**. Generally of greenish and often relatively impure material, they are almost never decorated. The body tends to be conical, pear-shaped, or oval; the neck is mostly cylindrical and has an outsplayed, folded rim. Since the basic form was probably introduced about the turn of the millennium and was still in use in the second and even the third centuries, it is most difficult to date specific objects by their shape alone. In general, however, this bottle form and its variants seem to have occurred most frequently in the first and second centuries. Literally tens of thousands were found in Palestine, Cyprus, Turkey, Egypt, Italy and the Rhineland, and in many other regions where the finds from archaeologically supervised excavations bear out this dating.[71]

At Sardis, more than two dozen bottles were found

under controlled conditions in datable graves[72] and excavation levels with fairly limited chronological boundaries. For this reason, the finds from graves and other precincts are recorded by findspot rather than grouped according to types, as in the previous sections. The dated bottles confirm the established chronology.

A number of bottles datable to the late first and early second century come from Ahlatlı Tepecik on the shore of the Gygean Lake (*BASOR* 191, 9). With the exception of **117**, they represent the simplest types of ordinary Roman glassware. A typical example is **110** with the conical body representing about one half of the total height.[73] Variations of such a form are shown by **111-116**.[74] Vessel **117** with conical neck similar to **135** appears to have close relatives in Cyprus and elsewhere.[75]

The bottles from the Roman graves at HoB, opened in 1959, belong to the late first to mid-second century (*BASOR* 157, 28). A coin (ca. A.D. 100-120) was found in one of the graves. Vessel **118** belongs to the group represented by **110**.

A fragmentary miniature bottle **120**[76] comes from a disturbed level at MTW and is probably early or mid-Roman Imperial. Two bottle necks found together at MTE, **121** and **122** may be of slightly later date (*BASOR* 177, 14-15).

The graves excavated in 1960 and 1963 at PC contained objects of the first and second centuries, **123, 127, 129-131** (*BASOR* 162, 18; 174, 23). A bottle with a very low body and a long conical neck, made of eggshell glass, **123**, comes from a burial of about A.D. 100.[77]

71. Isings, nos. 28, 82. Of the extensive material published cf. esp. Vessberg, *Cyprus,* 156ff; Harden, *Karanis,* nos. 797ff.; Davidson, *Corinth* XII nos. 666ff.; Clairmont, 130ff., nos. 671ff.; Lancel, 60ff.; Calvi, 28ff.; Edgar, pls. 7-8. Most of the material is datable to the 2nd C. The earliest pieces seem to begin not much later than the turn of the millennium, increasing in volume during the course of the 1st C.; the basic forms tend to taper off again in the 3rd C.

72. Grave contents at Sardis include relatively few glass bottles. In contrast, early Imperial burials in other regions within the Roman Empire have often revealed a great number. For example, some of the tombs at Minusio in the Ticino contained more than two dozen ordinary bottles or unguentaria with conical bodies and elongated necks (Simonett, 148ff., fig. 128, grave 14).

73. Isings, nos. 26, 28b. Cf. also Spartz, nos. 55ff.; Vessberg, *Cyprus,* figs. 48-50.

74. Cf. n.73 supra and Spartz, nos. 9ff.

75. Vessberg, *Cyprus,* fig. 48:32, 14. Kisa, *Glas* III, Formentaf. B, nos. 75ff.

76. Cf. n.74 supra.

77. For vessels with particularly wide, low body cf. Saldern et al., *Slg. Oppenländer,* no. 221 (with bibliographic ref.). Barag, "Glass Vessels," pl. 19, center. Eisen, fig. 189, pls. 66-67. A later develop-

Other bottles can also be dated safely to the second century. One, with spherical body, **124** was found with **126**, in grave 63.2 at PN; it is typologically close to **110** and **112** although it has a pronounced constriction at the base.[78] With these bottles came a coin of Antoninus Pius (A.D. 138–161; *BASOR* 174, 23). Spherical or pear-shaped bottles with conical necks such as **124** are known from other sites in the East as well as in the West and most are dated to the second century.[79]

A few other bottles and bottle fragments are datable either to the late first or second century. Bottles **127** and **128** belong to the same group as **110**, **112**, and **126** (*BASOR* 162, 18). A spherical bottle with conical neck, **129**, was discovered with a ceramic lamp of the mid–first through second century (L60.31:2709, red on white slip cf. Perlzweig, *Agora* VII no. 164) in tomb E at PC; the shape of the neck is reminiscent of that of **117**.

The following pieces appear to be of mid to late Roman date. An ovoid bottle with a neck showing a constriction at the base, **130**, and a fragment, **134**, from pit 1 in LVC at PC (*BASOR* 157, 14, 16) and may be third century.[80] Vessel **134** seems to represent a later development of the tall-necked bottles of the second century.[81] **132** was found in grave 63.1 at PN (*BASOR* 174, 22–23) with burial goods datable to the third or early fourth century. It had either a wide funnel-shaped mouth with folded rim characteristic of fourth century finds from Karanis,[82] or a relatively long neck with flaring, folded rim typical of bottles with spherical, double-walled body, i.e. having a base pushed up so high that it almost touches the wall of the upper body.[83] Bottles of this form are usually datable to the late second and third centuries.

Perhaps the most interesting bottle is **131** which is decorated with wheel incisions. It belongs to a well-known type common in both the East and West.[84] Most of the published examples are datable to the second half of the fourth century although this form seems to have been introduced by the mid-third century. Of the examples preserved a relatively large number have shallow grooves around body and neck; others have inscriptions engraved around the shoulder.[85] The bottle from Sardis was found in a burial pit of SVC at PC (*BASOR* 157, 14, 16). Both SVC and the LVC contained primarily third century material, including the portrait head of a priest[86] datable to the second half of the third century. Thus, **131** appears to be late third or possibly early fourth century.

Closely related to the small bottles from Ahlath Tepecik, **113–118**, is a miniature bottle from PN, **133**,

of uncertain date. It comes from a much disrupted level containing a system of waterpipes, probably Roman (at *89.00; *BASOR* 166, 16).

From grave 58.H at Kâgirlik Tepe[87] comes a bottle with conical neck, **135**, that is related to **117**; it was found with a ceramic lamp of the late first to second century (L58.34:784).

A rectangular bottle with brilliant iridescence is probably datable to the late first or early second century, **136**; found under a late Roman house, its exact date cannot be determined through stratification.[88]

Three bottle necks with rim folded outward, upward, and inward to form a sort of flange, come from unstratified levels, **137–139**. They may have belonged to pear-shaped or bulbous bottles with neck constriction. Bottles of this type occasionally bear applied threads of prunts[89] and usually occur in the third century. Another neck, **140**, is concave and shows a beveled rim. It comes from the disturbed area UT which contained much late Hellenistic and Roman material. Although its profile is not unlike that of **131**, the carefully worked and ground rim reminds one of early Imperial glass.

The square base, **141**, seems to be from a late first to second century bottle with handles.[90]

In 1968, thirty-six bottles were bought from a member of the village. Although their exact provenance is uncertain, their homogeneity and their similarity to the bottles described above make it practically certain that they are roughly of the same date and were probably made in the same region (or possibly even the same factory). Almost all of them are relatively badly made and of very thin material.

The first series, **142–155**, consists of bottles with a low, squat body, slender cylindrical neck, and wide

———
ment: Vessberg, *Cyprus*, fig. 60:6–7. Cf. also Harden, *Karanis*, 267, nos. 838ff., pl. 20.

78. Isings, no. 82A.

79. Ibid., no. 92, with Western examples of the 2nd C.

80. Cf. ibid., no. 103 and Vessberg, *Cyprus*, fig. 50.

81. Isings, no. 82A. Cf. Vessberg, *OpusArch*, pl. 9:4ff.

82. Harden, *Karanis*, 187, nos. 516ff., pl. 17. Cf. Isings, no. 133.

83. Spartz, no. 84 (with ref.).

84. Isings, no. 104.

85. Fremersdorf, *Gläser mit Schliff*, pl. 114; bottles of this shape rarely bear gilt decoration: ibid., pl. 274. Cf. also Saldern et al, *Slg. Oppenländer*, no. 519.

86. *Sardis* R2 (1978) no. 93, figs. 208–211.

87. *Sardis* R1 (1975) 127, 128.

88. Very similar: Clairmont, no. 757. For related types cf. Harden, *Karanis*, nos. 568–569; Spartz, nos. 49, 51.

89. *Smith Coll.*, nos. 315, 322, 326.

90. Isings, no. 50.

folded rim. In general, comparable finds date from the second century.[91]

Three bottles, **156–158**, have a more pear-shaped body. They belong to the same period.[92] A squatter variant is represented by **159**.

Another group includes bottles with a conical body that makes up about one-third of the total height of the vessel, **160–162**. Although listed by Isings as being of first century date, these bottles were also made in the second century.[93]

163 and **164** are almost identical to those of the previous series save for the height of the body which measures about one-half of the total height.[94]

The bottles **165–167**, derivatives of base forms, are closely related to the four groups just described. In most cases their shapes are the result of the glass-maker's whim rather than of changes in style. Practically all of them have forms included in Isings.[95]

Pieces **168** to **172** are transitional between those with a conical and those with a rounded, oval body.[96]

A miniature bottle, **173**, belongs in the same general category[97] while another bottle with oval body and high neck, **174**, is similar to a bottle found in LVC at PC, **130**, which is probably of third century date.

Finally, two conical bottles, **175** and **176**, with funnel neck and plain rim appear to belong to a slightly different group. They might be derivatives of a type described by Isings and could be dated to the mid- or late Imperial period.[98]

The problem of whether ordinary bottles of the types listed in this chapter were imported from regions along the Eastern Mediterranean or made in Asia Minor cannot be solved at present. Only the discovery of large numbers of stylistically and technically homogeneous glass vessels—such as the Early Byzantine material at Sardis—or the discovery of an actual factory site in Turkey could prove that Roman glass was also manufactured in Asia Minor. On the other hand, the presence of fairly numerous examples of early and mid-Imperial common ware in Sardis as well as other sites and the astonishingly extensive material on the antiquities market which stylistically and technically matches the excavated examples from Anatolia support the theory that glassmaking facilities must have existed somewhere (along the coast?) in Asia Minor by at least the late first and second centuries.

110 *Pls. 5, 21*. G67.10:7490. Conical with concave base, cylindrical neck with base constriction, infolded rim. Greenish; heavy iridescence and dark scum.

H. 11.7; diam. 5.5, of rim 3.8.
AhT grave 67.34, W7.62/S47.58–S47.67 *74.96. Found with **111**. Other finds in the grave were: an Eastern Sigillata B II bowl (P67.88:7486), A.D. 75–150; a lamp (L67.37:7491) of about the same date; ceramic piriform unguentaria (P67.87:4485 plus 2 uninventoried pieces), late 1st C. B.C.–1st C. A.D.; and a coin of the 1st–2nd C. A.D. under the skull.

111 G67.8:7488. Fragments of ampulla(?) with bulbous body and cylindrical neck. Greenish.
Diam. rim 1.8.
AhT grave 67.34, W7.62/S47.58–S47.67 *74.96. Found with **110**.

112 *Pls. 5, 21*. G67.11:7520. Conical with concave base, cylindrical neck, infolded rim. Aquamarine.
H. 10.6, diam. base 2.5.
AhT grave 67.32, W5.95–W6.40/S49.10–S49.50 *75.28–75.18. Among the grave finds were a terracotta figurine of Aphrodite (T67.14:7495); a carinated red ware bowl (P67.91:7496); 2 ceramic unguentaria (uninventoried); a mold-made lamp (L67.42:7515; cf. Perlzweig, *Agora* VII no. 1416); and 2 coins: one an autonomous Imperial, A.D. 54–68, and the other of the 1st–2nd C. A.D. Found with **113** and **114**.

113 *Pls. 5, 21*. Flattened oval miniature flask, infolded rim. Aquamarine; mended, small sections missing.
H. 5.7, max. diam. 2.5, min. 1.1.
AhT grave 67.32, W5.95–W6.40/S49.10–S49.50 *75.28–75.18. Found with **112** and **114**.

114 *Pls. 5, 21*. G67.9:7489. Miniature and spherical with tapering neck and infolded rim. Pale aquamarine.
H. 4.2, diam. 2.1.
AhT grave 67.32, W5.95–W6.40/S49.10–S49.50 *75.28–75.18. Found with **112** and **113**.

115 *Pls. 5, 21*. Spherical with pushed-up base, slender neck with infolded rim. Eggshell, clear with green tint; mended, small sections missing.
H. ca. 8.5, diam. rim 2.2.
AhT grave 67.37, E3.00/S52.40–S52.50 *76.23. Also in the

91. Ibid., no. 82:A2, B2, here called candlestick unguentarium.
92. Ibid., no. 82:A1.
93. Ibid., no. 28b.
94. Ibid., no. 28a.
95. Ibid., nos. 8, 28, and 82. Cf. also Vessberg, *OpusArch*, 136ff. Harden, *Karanis*, 265ff. For **165** cf. Isings, no. 28a (without constriction at the base of the neck). For **166**, a variant with oval body, cf. Vessberg, *OpusArch*, pl. 9:5ff. For **167**, a variant with spherical body, cf. Vessberg, *OpusArch*, pl. 8:16.
96. Vessberg, *OpusArch*, pl. 9.
97. Ibid., pl. 9:23.
98. Isings, no. 28a.

grave were: an Eastern Sigillata B I footless bowl with everted rim (P67.114:7539), 1st C. A.D., and a Julio-Claudian coin under the skull.

116 *Pls. 5, 21.* G67.7:7470. Miniature, bulbous with short neck. Eggshell, greenish; heavy iridescence, dirty scum; part of lip missing.
H. 4.7, diam. 2.9.
AhT grave 67.28, E0.40/S49.60 ca. *76.66. Among the grave finds were ceramic piriform unguentaria of the 1st C. B.C.–2nd C. A.D.; Eastern Sigillata B I pottery (P67.73:7453 and uninventoried); and a coin attributed to Augustus-Tiberius, late 1st C. B.C.–early 1st C. A.D.

117 *Pl. 5.* Concave base, probably pear-shaped body, funnel neck. Eggshell, aquamarine; broken in many pieces, large sections of body missing.
Est. H. ca. 10.0; diam. base ca. 5.0, rim 4.5.
AhT grave 67.18, W10.30/S52.20-S52.25 *74.97 (top). Found in the grave were 2 coins (one in each of the 2 skulls) of the 1st C. B.C.–2nd C. A.D.

118 *Pl. 5.* G59.47a:1851. Hemispherical body with concave base, slightly conical neck. Greenish; rim broken.
H. 19.2, diam. 9.0.
HoB grave 59.h, E20/S60 *98.68.

119 *Pl. 6.* G59.45a:1843. Spherical with slight kick, cylindrical neck, infolded rim. Thin, clear with green tint.
H. 7.6.
HoB grave 59.g, E10/S60.
AJA 66, 8, no. 9a, pl. 6:8.

120 *Pl. 21.* Miniature, conical with high kick, cylindrical fractured neck. Greenish; heavy black scum.
P.H. 5.7, diam. 2.2.
MTW 1964 ca. W0-W8/S210-S215 to ca. *116.20.

121 Irregular short neck with infolded rim, increasing in diam. towards (cylindrical?) body. Greenish; heavy brown scum.
P.H. 3.0, diam. rim 2.3.
MTE 1964 E50-E55/S140-S143 to *101.70. Found with **122**.

122 Fragmentary cylindrical neck widening to irregular, infolded rim. Green tint.
P.H. 4.0, diam. rim 3.2.
MTE 1964 E50-E55/S140-S143 to *101.70. Found with **121**.

123 *Pls. 6, 21.* G60.31:2736. Part of upper body and whole neck, squat ovoid body, slender neck tapering downwards, base ring. Eggshell, clear, green-blue tint; slight decomposition.
P.H. ca. 16.0, H. neck 12.0; diam. body 11.0, base ring 3.8.
PC tomb C *91.50.
AJA 66, 9, no. 9f, pl. 6:12.

124 *Pl. 6.* G63.16:5407. Spherical, concave base, slightly conical neck. Eggshell, clear with green tint.
H. 12.8; diam. 8.2, rim 3.0.
PN grave 63.2, W228/S341 *87.90.
BASOR 174, 23.

125 *Pl. 21.* G59.39:1472. Bottom part of globular (hemispherical?) body with concave base. Aquamarine.
Max. dim. 17.0, est. diam. 20.0.
HoB E10/S60, from fill, *99.40-99.00.

126 *Pl. 6.* G63.15:5376. Conical with kick, cylindrical neck with base constriction, infolded rim. Greenish.
H. 14.3, diam. base 3.0.
PN grave 63.2 within sarcophagus, W228/S341 ca. *88.40-87.85.
BASOR 174, 23.

127 *Pl. 6.* G60.27:2732. Conical with concave base, cylindrical neck with base constriction. Aquamarine; upper neck missing.
P.H. 8.5; diam. base 2.3.
PC, E of LVC *90.50.
BASOR 162, 18, n.19.

128 *Pl. 6.* G60.25:2730. Cylindrical neck with base constriction, flattened infolded rim (carefully made). Aquamarine.
P.H. 9.7, diam. rim 2.9.
PC zone A *93.14-92.00.

129 *Pls. 7, 21.* G60.32:2742. Spherical with concave base, conical neck. Eggshell, clear with green tint.
P.H. 12.2.
PC tomb E *91.00.
AJA 66, 9, no. 9e, pl. 9:9e.

130 *Pl. 7.* G59.46B:1850. Ovoid body with kick, slender cylindrical neck with base constriction, infolded rim. Thin, clear with green tint.
H. 15.7.
PC LVC, pit 1.
AJA 66, 8, no. 9b, pls. 6:9, 8:9b.

131 *Pls. 7, 22.* G59.48a:1855. Spherical, foot ring, neck flaring to plain rim; 4 shallow cut grooves around body, 2 around upper neck. Clear, green tint.
H. 10.8; diam. body 8.1, rim 3.8.
PC SVC, pit 2.
AJA 66, 8–9, no. 9d, pls. 6:11, 9:9d.

132 *Pls. 7, 22.* G63.6:5234. Bell-shaped with kick, cylindrical neck with base constriction. Aquamarine; partly mended, upper neck with wide folded rim missing.
P.H. 13.0, diam. 6.0.
PN grave 63.1, W222/S348 *89.70, beside skeleton in sarcophagus.
BASOR 174, 22ff.

133 *Pl. 7*. G61.8:3196. Hemispherical body with kick, slender cylindrical neck with folded rim. Yellowish, bubbly.
H. 7.1.
PN W245/S375 ca. *89.40.

134 *Pl. 7*. G59.46A:1850. Globular with pushed-up base, cylindrical neck. Greenish; rim missing.
P.H. 18.5, diam. 8.5.
PC LVC, pit 1.

135 *Pl. 22*. G58.14:420. Oval with slight kick, conical neck. Aquamarine, eggshell; fractured.
H. 11.5; diam. 8.0, diam. rim 3.0.
KG grave 58.H *187.15-186.46 a.s.l. (1.20–1.45 m. below datum).
Hanfmann, *JGS* 53; *BASOR* 154, 13; *Sardis* R1 (1975) 126ff., fig. 328.

136 *Pls. 7, 22*. G60.1:2242. Four-sided (each side slightly indented), concave neck, infolded rim. Probably clear, green tint; brilliant iridescence.
H. 6.0.
PC (zone Z) area of early Roman tiled graves, under late Roman house *91.00.
AJA 66, 8, no. 9c, pls. 6:10, 8:9c.

137 *Pl. 22*. Cylindrical neck with base constriction, widening to rounded shoulder, wide, out- and under-folded rim. Aquamarine; hole at side of neck.
P.H. 5.3, diam. rim 4.0.
MTE 1964 E67-E72/S157-S161 *111.90-111.70.

138 Upper portion of cylindrical neck, wide, out- and under-folded rim. Green tint.
Diam. 3.6.
HoB 1963 W20/S110-S115 to *100.80.

139 Fragment of neck, see **138**. Green tint.
Diam. 4.2.
MTE 1964 E64-E72/S155-S161 to *112.20.

140 *Pl. 22*. Concave neck, widening to body, beveled edge. Aquamarine; flaking iridescence; well made.
P.H. 3.5, diam. 3.8.
UT 1959 E70/S200 to *119.68.

141 *Pl. 22*. Concave base fragment of square bottle; square in cross section. Olive; scaling iridescence.
Est. W. of one side ca. 5.6.
HoB 1961 E5/S90-S95 *99.80-99.60.

142 *Pl. 8*. NoEx 68.17.31. Squat conical, cylindrical neck, outsplayed infolded rim. Aquamarine.
H. 9.0, diam. base 4.9.

143 NoEx 68.17.13. Form as **142**, thick rim. Clear; brown rough scum.
H. 8.3, diam. base 5.0.

144 NoEx 68.17.17. Form as **142**, rough pontil mark, irregular rim. Aquamarine; rough scum.
H. 10.0, diam. base 4.8.

145 NoEx 68.17.6. Form as **142**, high kick, infolded irregular rim. Bluish green tint; frosted, brown decomposition.
H. 11.7, diam. base 5.7.

146 NoEx 68.17.19. Form as **142**, neck constricted, infolded rim. Aquamarine; frosted brown decomposition.
H. 11.2, diam. base 4.5.

147 NoEx 68.17.24. Form as **142**, high kick and rough pontil mark, infolded irregular rim. Green tint.
H. 14.5, diam. base 5.5.

148 NoEx 68.17.2. Form as **142**. Concave base, infolded rim. Green tint; white corrosion.
H. 11.3, diam. base 7.5.

149 NoEx 68.17.30. Form as **142**, high kick, infolded rim. Clear, bubbly.
H. 15.4, diam. base 6.2.

150 NoEx 68.17.26. Form as **142**, constricted neck, infolded irregular rim. Clear.
H. 16.5, diam. base 6.2.

151 NoEx 68.17.16. Form as **142**, slight kick, rough pontil mark, constricted neck, infolded irregular rim. Green tint, bubbly.
H. 17.0, diam. base 7.4.

152 NoEx 68.17.1. Form as **142**, rough pontil mark, constricted neck, infolded irregular rim. Aquamarine tint.
H. 17.8, diam. base 8.0.

153 NoEx 68.17.34. Form as **142**, but regular, concave base, infolded rim. Aquamarine.
H. 17.0, diam. base 7.8.

154 NoEx 68.17.27. Form as **142**, concave base, constricted neck, infolded rim. Bluish green tint; white corrosion.
H. 11.0, diam. base 5.5.

155 *Pl. 8*. NoEx 68.17.18. Squat conical, concave base, cylindrical neck, infolded irregular rim. Bluish green tint; brown decomposition; mended.
H. 12.3, diam. base 6.4.

156 *Pl. 8*. NoEx 68.17.3. Pear-shaped, cylindrical neck, infolded irregular rim. Bluish green tint; rough scum, corroded.
H. 9.4, diam. base 3.8.

157 NoEx 68.17.33. Form as **156**, high kick, constricted neck, infolded rim. Yellow tint; corroded brown scum.
H. 15.7, diam. body 6.3.

158 NoEx 68.17.4. Form similar to **156**, high kick. Bluish green tint, bubbly; corroded, brown scum.
H. 11.3, diam. body 4.0.

159 *Pl. 8*. NoEx 68.17.5. Hemispherical, concave base, constricted cylindrical neck, infolded rim. Aquamarine; rim fractured.
H. 18.3, diam. base 9.9.

160 *Pl. 8*. NoEx 68.17.29. Conical, rough pontil mark, cylindrical neck. Bluish green tint, bubbly; brown decomposition; rim missing.
H. 14.5, diam. base 3.5.

161 NoEx 68.17.10. Form as **160**. Bluish green tint; brown decomposition; rim missing.
H. 13.2, diam. base 3.2.

162 NoEx 68.17.21. Form as **160**, concave base, infolded rim. Aquamarine; corroded, brown scum; fractured.
H. 14.0, diam. base 3.6.

163 *Pl. 8*. NoEx 68.17.8. Conical, rough pontil mark, cylindrical neck, infolded rim. Aquamarine, eggshell; dull, brown decomposition.
H. 10.0, diam. base 3.7.

164 NoEx 68.17.25. Form as **163**, infolded irregular rim. Aquamarine, large bubbles.
H. 8.8, diam. base 3.5.

165 *Pl. 8*. NoEx 68.17.15. Conical, cylindrical neck, infolded rim. Greenish; brown decomposition.
H. 7.0, diam. base 4.4.

166 *Pl. 8*. NoEx 68.17.35. Oval, cylindrical neck, infolded irregular rim. Green tint; frosted, brown scum.
H. 10.5, diam. body 3.5.

167 *Pl. 9*. NoEx 68.17.9. Irregular, spherical with kick, cylindrical neck, infolded rim. Green tint, bubbly; brown scum.
H. 9.7, diam. body 3.0.

168 *Pl. 9*. NoEx 68.17.12. Conical, cylindrical neck, outsplayed rim. Clear, eggshell.
H. 10.2, diam. base 3.8.

169 *Pl. 9*. NoEx 68.17.36. Almost conical, cylindrical neck, infolded rim. Clear, eggshell; brown decomposition.
H. 11.3, diam. base 4.2.

170 *Pl. 9*. NoEx 68.17.7. Hemispherical, slender neck, infolded rim. Clear, eggshell; brown decomposition.
H. 7.5, diam. base 3.5.

171 NoEx 68.17.11. Form as **170**, high kick, infolded, irregular rim. Bluish green tint, eggshell; frosted, dark scum; rim fractured, hole at body.
H. 10.8, diam. body 4.5.

172 *Pl. 9*. NoEx 68.17.23. Cylindrical neck, outsplayed, infolded, and irregular rim. Clear, eggshell.
H. 11.3, diam. base 3.2.

173 *Pl. 9*. NoEx 68.17.28. Oval, neck tapering upward. Bluish tint, eggshell.
H. 4.6, diam. body 1.7.

174 *Pl. 9*. NoEx 68.17.32. Oval, high kick, constricted cylindrical neck, infolded irregular rim. Green tint, bubbly.
H. 17.3, diam. body 5.6.

175 *Pl. 9*. NoEx 68.17.22. High conical, slight kick, tapering neck. Clear; white corrosion; mended.
H. 18.3, diam. base 7.7.

176 NoEx 68.17.14. Same as **175**.
H. 19.0, diam. base 7.5.

VESSELS WITH STRAP HANDLES

Because of their thickness, vessel handles and bases were relatively sturdy and, therefore, tended to survive intact more frequently than the fragile vessel walls. Even a small portion of a handle will give some information on the type of the original vessel, while a wall section of the same size is usually much less informative.

At Sardis, a series of corrugated strap handles, with tooled grooves running vertically on the exterior, were found at various locations. Most of them appear to belong to well-known groups of early to middle Roman Imperial vessels, primarily squat or high bottles with single or double strap handles (*Pl. 10*). According to the archaeological evidence, however, one or more may date to late or even post-Roman times.[99]

177 was discovered at PN in a level dated by a coin of Valentinian I (A.D. 364–375; *BASOR* 170, 20). Fragments **178–180** found at HoB belong to mixed Roman occupation levels. **179** was from the bottom portion of a vessel—perhaps a beaker—with concave base.[100] **180** was excavated with ordinary aquamarine-colored window glass. **181** was found under BS W 13 with late Hellenistic and early Roman sigillate ware (*BASOR* 157, 34); in this case we can be fairly certain

99. Isings, nos. 50ff. Spartz, nos. 46ff. (with ref.).

100. Such bases could have belonged to beakers, cf. Vessberg, *OpusArch,* pl. 3:29; Isings, no. 32.

that the handle is datable to about the first century A.D. A handle found in BS W 2, **182**, is possibly fourth century.[101] Finally a handle from BE–S, **183** was found with a small single-thread handle of a bowl or lamp of Early Byzantine date, apparently an indication that vessels with strap handles were also made after the fourth century.

177 *Pl. 22.* Portion of handle and rim section of jug or bottle with plate-like mouth, infolded rim and handle with narrow, shallow ribbing. Clear with green tint.
P.H. 4.5, W. 3.7, diam. rim 4.6.
PN 1962 ca. W220-W240/S365-S385 *89.45.

178 *Pl. 22.* Lower part of strap handle from bottle or jug with portion of wall from shoulder; 5 pronounced ribs. Bottle green, bubbly.
P.H. 7.0, W. 5.0.
HoB 1965 W15-W25/S125-S128.80 to *101.85-101.60.

179 Portion of corrugated handle. Greenish.
P.H. 2.5.
HoB 1962 E5-E10/S115 to *100.60.

180 Portion of handle with corrugation deeper than **179**. Aquamarine.
P.H. 3.0.
HoB 1962 W10/S115 *101.00.

181 Large portion of handle with narrow corrugations. Aquamarine.
P.H. 4.5, W. 3.0.
BS W 13, 1959 W54-W57/S2-S4, pit below floor in SW corner, *94.50-94.00.

182 G58.39:547. Portion of handle. Aquamarine.
P.H. 5.8.
BS W 2, W7.5/S0-S3 level II, below top of S wall.
Hanfmann, *JGS*, 53, n.11.

183 Flat, corrugated handle. Greenish.
BE–S 1965 E26.75-E29/N27-N36.25 to *96.10.

184 Section of handle(?), 2 trunk-like applications with portions of curved walls from a vessel of unidentified shape. Green tint, eggshell; white scum.
P.H. 3.2.
HoB 1965 unstratified.

VESSELS WITH FLAT BASES (BEAKERS OR BOWLS)

A series of base fragments appears to belong to a specific vessel type. The base is flat and set off against the outsplayed, lower portion of the vessel which, in analogy to bowl-like beakers with the same base

(found, for example, in Dura Europos and Cyprus), seems to have been of roughly hemispherical form with the rim turned outward.[102]

The fragments from Sardis are greenish and are now covered with a whitish, enamel-like decomposition not unlike the cut fragments **64–76** and those with warts, **81–90**. Two of the sherds, **185**, were found in the area where ca. 150 nondescript vessel fragments of greenish and aquamarine glass of probably mid-Imperial date were found, including the section of a late second or third century faceted bowl, **65**. The third fragment, **186**, is unstratified while a fourth, **187**, was found together with the fragment of a vessel with warts, **81**, also datable to the late second or third century.

As these approximate dates concur with the period to which the beakers found in Dura and in Cyprus have been assigned, the latter part of the second and the third century, it seems highly likely that the vessel bases recorded here also date from this time. In view of the parallel material it is equally likely that they formed part of bowl-shaped beakers (n. 102, supra).

185 *Pl. 22.* Two base sections of cups or beakers; flat base set off against convex wall. Clear with green tint; heavy silver white iridescent scum.
P.H. 5.0 and 4.5, diam. base 5.0, Th. base 0.7.
MTE 1964 E60-E65/S149-S153 *110.00-109.50.

186 Small base fragment (cf. **185**). Clear with green tint.
HoB 1962 E5-E10/S115 to *101.40. Found with **201** and **582**.

187 *Pl. 10, 22.* Same as **186**.
Diam. 4.8.
UT 1959 E80/S200 *121.50-121.00. Found with **81**.

BOWLS WITH PRONOUNCED BASE RINGS

A common type of Roman glass is a fairly deep bowl with convex, conical, or slightly flaring walls, resting on a sloping base ring.[103] A few sections of base rings which apparently belonged to such bowls were found in Sardis. All examples recorded have been knocked off very close to the base, leaving fractured edges which give no substantial information

101. Cf. Hanfmann, *JGS*, 53, n.11.
102. Clairmont, 97, nos. 432ff., pl. X. Vessberg, *OpusArch*, 122–123, pl. 3:37–39. Cf. also Loeschcke, *Slg. Niessen*, pl. 23:274 (with a row of slight indentations, here *Pl. 22*).
103. Cf. for example, Harden, *Karanis*, nos. 221ff., 304, pl. 14 (nos. 221, 304, here *Pl. 22*). Clairmont, nos. 76ff., 95ff., pls. 2, 3; nos. 384ff., pl. 9 (bases only).

about the profile of the complete vessel. Only an analysis of the profile and the diameter of the base ring will contribute to a reconstruction of the original vessel type.

The base rings from Sardis are sloping and either straight-sided or curved slightly toward the exterior. The bottom of the vessel is generally slightly convex and extends beyond the base ring in an almost horizontal direction. Such a profile may be typical of a variety of bowls and plates dating from early Roman to late Imperial times.[104] But it seems most likely that these base rings were from bowls of the general type referred to in note 103. The character of the fabric and decomposition, the average diameter of the bases, as well as the fact that most of the Roman glass finds from Sardis tend to be of late first to third century dates appear to bear out this supposition. **188** was found in a grave at PN that appears to be of third or early fourth century date (*BASOR* 174, 22–23).

The other sherds from PN, **189–191**, come from mixed Roman levels (*BASOR* 174, 20ff.; 177, 3–4); the base ring from HoB, **192**, was in much disturbed, fairly high fill over the Lydian levels (*BASOR* 166, 5–7).

188 *Pl. 22.* G63.8:5253. Complete base ring of bowl(?), sloping downward. Probably green tint; heavy brown scum, silver iridescence.
Diam. 6.0, H. of ring 1.5.
PN grave 63.1, W221/S347 ca. *88.70-88.60.

189 Section of base ring. Green tint; irregular.
Est. diam. 4.6.
PN 1963 W241/S356 *90.00.

190 Section of base ring. Green tint.
Est. diam. 6.0.
PN 1964 W254-W255/S339-S340 *88.20-88.10.

191 *Pl. 22.* Section of base ring. Green tint; irregular.
Est. diam. 6.0.
PN 1964 W254-W256/S342-S345 *87.45-87.15.

192 *Pl. 22.* Section of base ring, fractured edge of broken off vessel at top. Green tint; heavy silver scum.
P.H. 5.0, est. diam. 9.0.
HoB 1961 W10/S90-S95 to *101.15.

VARIOUS VESSEL BASES

This and the following groups of vessel fragments include a great variety of small sherds most of which are very difficult to associate with specific vessel forms. They appear to date from mid-Imperial times.

A base, **193**, with pushed-up bottom and base ring, was found in a pit in LVC at PC (*BASOR* 157, 15–16); it could have belonged to a large, hemispherical bottle with cylindrical or conical neck, a shape popular in the third century.[105] This base was discovered with a number of nondescript fragments, representing about five vessels, perhaps bottles or bowls. The next base, **194**, also with a base ring, could have belonged to a small bottle or bowl of mid- or late Roman (or Early Byzantine?) date; the sherd seems to belong to a level over early Roman strata (*BASOR* 162, 17ff.). **195** and **196** cannot be assigned to any clearly definable period as they come from the disturbed levels at UT. **195** is the lower portion of a vessel with slightly concave, waisted base, and walls that are faintly conical at the lower end. Whether the original object was a bowl or a bottle cannot be determined. The shallow, lower part of a vessel, **196**, is perhaps from a bottle, which has an applied base ring in the form of a two-strand spiral coil. A flat base of very common construction, **197**, was found at a high level at HoB (*BASOR* 174, 6); it could either be late Roman or Early Byzantine.

Two vessel portions that appear to be from goblets, **198** and **199**, may be mid- or late Roman (or even Early Byzantine). A sloping foot and lower part of a bowl that is egg-shaped at bottom comprise **198**; it comes from a high level at HoB.[106]

193 Lower portion of vessel with convex walls and straight base ring. Greenish.
Est. diam. ca. 10.0.
PC LVC 1959 ca. *93.40-93.35.

194 G60.26:2731. Lower section of vessel (cup?) with convex walls, straight-sided base ring. Green tint.
Diam. 3.3, Th. to eggshell.
PC zone A *93.14-92.00.

195 Slightly concave base section of cup(?), spreading wall. Almost clear.
Est. diam. 5.0.
UT 1959 wall trench, E50/S230-S235 *125.30-124.50.

---

104. Cf. 1st C. skyphoi with circular base rings, Isings, no. 39, or the relatively late, i.e. 3rd–4th C. dishes and bowls, Harden, *Karanis*, pls. 11, 14.

105. Cf. for example, Harden, *Karanis*, no. 638, pl. 18, showing a very similar base construction.

106. It could have come from a goblet similar to pieces such as E. B. Dusenbery, "Ancient Glass in the Collection of Wheaton College," *JGS* 13 (1971) 22, no. 35, fig. 34; Saldern et al., *Slg. Oppenländer*, no. 706. Cf. also Isings, no. 111.

196 *Pl. 22*. Lower, shallow portion of vessel, probably a bowl; applied base ring made into a two-strand spiral. Aquamarine.
Diam. 7.0, Th. up to 0.15, W. of coil 1.0.
UT 1959 N side E80/S210 *122.65-122.10.

197 One of a group of fragments of thick clear glass with green tint, from vessels of unidentified shapes; all have heavy silver iridescence and frosting. All or most seem to come from similar cups.
Base fragment; flat base with slightly projecting edge, gently curving wall.
P.H. 4.0, est. diam. base 6.0.
HoB 1963 W25/S110-S115 *101.40.

198 Irregular, sloping foot, concave stem, egg-shaped lower bowl curving to cylindrical or slightly convex upper portion of the vessel (goblet?).
P.H. 2.5, diam. foot 3.1.
HoB 1962 W30/S90-S95 to *101.20.

199 Portion of lower (egg-shaped?) bowl, concave stem and upper portion of the foot. Greenish.
P.H. 3.4, diam. at constriction (stem) 0.9.
HoB 1965, chance find.

VESSELS WITH CURVED RIMS

At least three fragments, **200–202**, come from early Imperial beakers or bowls with walls whose upper portion is perpendicular to an out-turned, upward-curving rim.[107] **202** is decorated with faintly engraved lines often found on similar beakers.[108] The early Roman date of **200** is confirmed by the stratigraphic evidence; it was found in a low Roman level (*BASOR* 174, 13).

The other fragments are unstratified; **201** (*BASOR* 170, 13) was discovered together with the upper portion of a bottle with wide rim and waisted neck that is almost certainly Early Byzantine, **582**. Another rim fragment, **203**, very similar to those just mentioned may be from a similar beaker or bowl; the area where it was found has not as yet been thoroughly investigated (*BASOR* 174, 7–8).

200 *Pl. 10*. G63.4:5130. Perpendicular wall fragment of cup with rim bent out and up, ground edge. Aquamarine.
P.H. 6.0, est. diam. 9.0, Th. 0.25.
HoB W30/S100-S105 *100.20-99.90.

201 *Pl. 22*. Form as **200**. Bottle olive.
P.H. 6.0, est. diam. 8.0, Th. 0.1-0.2.
HoB 1962 E5-E10/S115 to *101.40. Found with **186** and **582**.

202 Rim fragment of bowl, slightly convex wall, ground rim turned outward and upward; faint cut lines, a wider line 4.5 below edge. Blue.
P.H. 6.0; est. diam. rim 6.5, body 7.0; Th. 0.2.
Chance find.

203 Outward and upward-curving rim fragment of bowl. Aquamarine.
P.H. 3.0, est. diam. 6.0, Th. 0.2.
HoB 1963 W13-W18/S120-S125 to *102.30.

MISCELLANEOUS VESSEL SECTIONS

The vessel fragments included here will not add much to our knowledge of the typology of Roman glass at Sardis. They include six rim fragments of vessels of unknown shape and one handle. The following commentary does not attempt to put them in their proper context (which is probably impossible).

A plain in-turned rim of a vessel, **204**, is undated; it was found with many nondescript fragments at MTE. It is significant because it is made of blue glass, a very rare color at Sardis.

Another fragment of a vessel, **205**, with plain rim waisted below the edge comes from a disturbed level at HoB (*BASOR* 166, 7ff.). The rim section (of a cup or bowl?), having no known provenance, **206**, seems to be related to early Imperial material despite its sprung rim, cf. **15–24**. A more complete rim section, **207**, is the upper section of a bottle or jar with its rim turned inward. It was found at MTE and is undated; it could be either late Roman or Early Byzantine (*BASOR* 177, 14). The glass fragments found with this rim section range from Roman vessels with cut grooves to Early Byzantine cups (or bottles) of the type represented by **401–444**.

The section of a bowl or jar, **208**, with a rope handle attached to the edge, also found at MTE, might be attributed to late Roman or Early Byzantine times; the former is more likely. From about the same location comes an angled handle with thumb rest, **209**, which appears to be Roman rather than Byzantine.[109] An unidentifiable, handle-like sherd, **210**, is also unstratified.

Among the other fragments is the section, **211**, of a cube-like piece of thick glass perhaps used for inlay work, showing an edge with a right angle. Two con-

---

107. Vessberg, *OpusArch*, 119, pl. 3:9ff.
108. Isings, no. 12.
109. It could be compared with similar handles of skyphoi, La Baume, no. L 4, pl. 49; *Antike und byzantinische Kleinkunst*, sale Helbing (Munich, Oct. 28–30, 1913) no. 788, pl. 28.

verging grooves are engraved on the top. The date is uncertain (*BASOR* 166, 48–49).

The identification of a wall fragment, **212**, is equally difficult. On top there are portions of two applied droplets that do not appear to have served as decorative devices but seem to be the remains of a broken-off section. The sherd is early Roman Imperial at the earliest. It was found at *96.10 in the Lydian Trench at HoB in a Geometric-Protogeometric context (*BASOR* 177, 13–14) but is clearly an intrusion such as might fall from the scarp during excavation.

Finally, two wall fragments of purple glass are worth mentioning because of the rarity of this color at Sardis and elsewhere. One, **213**, comes from PN and is certainly Roman (*BASOR* 182, 25). The other, **214**, found with a section of a rod, is from a fairly well stratified location west of the Marble Monument at PN at a level (*89.90-89.40) that is early Roman Imperial (first half of the first century(?); *BASOR* 174, 22–23). The piece might, therefore, belong to the group listed under **15–24**.

204 *Pl. 22*. One-third of vessel rim with edge turned inward, the rim being tooled but not folded. Light blue, eggshell.
P.H. 3.8, est. diam. 5.0.
MTE 1964 E60-E65/S149-S153 *110.00-109.50.

205 *Pl. 22*. Plain rim waisted below the edge. Aquamarine, eggshell.
P.H. 3.5, est. diam. 10.0.
HoB 1961 E5/S90 to *99.20.

206 *Pl. 22*. Rim fragment of shallow cup, sprung rim. Clear, yellowish green tint.
P.H. 6.5, est. diam. 16.0.
Chance find 1964.

207 *Pl. 22*. One-half of folded rim of bottle or jar, the edge turned outward. Greenish?
Diam. 5.0, Th. 0.2.
MTE 1964 E65-E70/S155-S160 *109.00.

208 *Pl. 22*. One-sixth of slightly flaring rim of bowl with waisted neck; handle-section twisted like a rope. Aquamarine.
P.H. 3.3, est. diam. ca. 7.0-10.0, Th. to 0.1.
MTE 1964 E65-E70/S155-S160 *109.00-108.50.

209 *Pl. 22*. Small handle bent to an almost r. angle, with small thumb rest at top. Greenish?
H. 6.0.
MTE 1964, upper SW area, 1.7 m. below walls.

210 *Pl. 22*. Two trunk-like applications (section of handle?) with portions of curved walls from vessel of unidentified shape. Green tint, eggshell; white scum.
P.H. 3.2.
HoB 1965 unstratified.

211 Section of a cube (used for inlay?); on top 2 cut, converging grooves. Clear with green tint.
P.H. 2.5, Th. 1.0.
BE-N 1961.

212 Rim fragment with remains of 2 irregularly shaped droplets (which appear to be portions of larger broken-off sections). Clear, yellow tint.
P.H. 3.5.
HoB 1964 W21-W25/S117.50-S119 *96.10.

213 Small section (from bottom?) of vessel. Purple.
Max. dim. 4.5, Th. 0.1-0.25.
PN 1965 W290-W295/S335-S340 *86.50-86.00.

214 G63.3B:5114. Slightly curved section of vessel. Deep purple.
Max. dim. 3.7.
PN W225/S345 *89.90-89.40.

STAMPS

Four stamps came to light at Sardis, three of which were unstratified. However, they are undoubtedly of Roman date.[110] One has a stamped impression of a Medusa head, **215**. The next, **216**, shows the figure of a bearded man in front of an altar (Lydian Zeus?) engraved on the surface. The third stamp, **217**, is decorated with the helmeted bust of a male (Zeus? Ares?), apparently impressed onto the surface. The fourth stamp, **218**, from B at a level associated with the Early Byzantine period, shows a man on top of two deer.[111]

215 *Pls. 10, 22*. G64.2:6129. Oval with head of Medusa; design very "soft" but seems to have been engraved. Greenish.
1.2 by 1.0, Th. 0.3.
MTE E69-E70/S150-S160 *109.00.

216 *Pl. 10*. NoEx 68.5. Oval with cut design: figure of bearded man standing and turned to r. (Lydian Zeus?) with curly hair and in tunic; his l. arm is held up over an altar

---

110. Kisa, *Glas* II 481; Harden, *Karanis,* 297–298 (for their use as merchandise marks, tokens, tickets, amulets, etc.).

111. The stamps will also be published with gemstones and other small finds by R. S. Thomas in a forthcoming Sardis monograph.

with flame (offering scene?); behind him a torch-like object. Aquamarine.
1.3 by 1.1.
Field W of HoB, chance find.

217 *Pl. 10*. Oval, with helmeted male bust in profile, turned 1. (Ares?); design recognizable only with help of the wax impression.
1.3 by 1.0.
Chance find 1964.

218 *Pl. 22*. Seal 58.1:986. Oval; man on top of (astride?) 2 deer. Greenish.
H. 1.7.
BS W 2 level II.

MISCELLANEOUS OBJECTS

Spiral Rods

A well-known group of Roman glass objects consists of spirally twisted and plain rods (similar to modern martini stirrers). Sometimes these objects reach considerable length—20 to 30 cm.—and often terminate at one end in a flat disc, a spoon-like motif, or a bird. Usually assigned to the early or mid-Roman period, they are rather common in the Eastern Mediterranean as well as in the West.[112]

Two spiral rod sections are safely dated to the early Imperial period. **219** was found in the vaulted passage beneath the Marble Court (*BASOR* 177, 26) with small vessel fragments of Roman date and **220** came from a level below a second or third century stratum (*BASOR* 174, 13). **221** comes from the Hellenistic grave k at HoB (*BASOR* 157, 28) but rods of this type have, to my knowledge, not previously been associated with the Hellenistic period. The piece could be an intrusion or belong to a time when glassmaking began to take its course on a relatively large scale, and objects such as rods and pins were among the standard products of a late Hellenistic workshop. **222** and **223**, come from unstratified levels.

Pins and Tubes

Sections of a few plain, very fragmentary pins or rods were found.[113] Used for stirring or as spatulae, they were certainly made over a long period of time; no doubt Roman examples will not have differed markedly from later examples.

A pin section of dark (blue?) glass was found at PN, **224**. Another section, of blue glass, **225**, comes from a mixed Hellenistic-Roman level at HoB (*BASOR* 166,

5–7). An almost identical rod section was found in an unstratified level at the Synagogue (unlisted). All these pieces resemble rods or needles of early Roman date found in northern Italy and elsewhere in the West[114] as well as in Syria.[115]

A pinhead, **226**, no doubt used for a metal pin, was found in a level that contained Hellenistic coins (for example: Philip III, 319 B.C.; Antiochus II, 261–246 B.C.; and Antiochus III, 213–190 B.C.) and Roman material (*BASOR* 177, 3–4); it may be tentatively assigned to the early Roman period.

Pendants

A bead-like pendant, **229**, unfortunately unstratified, shows decomposition comparable to that on Roman vessel fragments with applied warts, **81–90**.

A tear-shaped object, **230**, found in a Roman grave of the third or fourth century (*BASOR* 174, 22–23) resembles "drippings" or tiny bits of glass that fall off during manufacture and are usually found at factory sites. This piece, however, seems to have been fashioned into a pendant by adding a tiny loop at the top.

Ring

The only ring found at Sardis, **231**, appears to be Roman, and not Early Byzantine, although it is unstratified. It is of a simple type that is almost impossible to date accurately.[116]

Bracelets

Bracelets of Roman date are as rare as rings. Sections of only two bracelets were discovered that seem to be of pre-Byzantine date.[117] One, **232**, belonged to a Hellenistic-Roman fill overlaying Lydian levels

112. Isings, no. 79 (only Western examples). Harden, *Karanis*, 285–286, nos. 860–864, pl. 21. Vessberg, *Cyprus,* 212, fig. 51:15–17. Saldern et al., *Slg. Oppenländer,* nos. 619ff. *Smith Coll.,* nos. 298, 300. *JGS* 11 (1969) 111, no. 10. No less than 10 rods with lengths up to 24 cm. were found in a 1st C. grave (no. 7) at Muralto in the Ticino: Simonett, 75. Cf. also Kisa, *Glas* II 353ff.

113. Harden, *Karanis,* 286–287. Fremersdorf, *Buntglas,* pl. 124.

114. Fremersdorf, *Buntglas,* pl. 124. Saldern et al., *Slg. Oppenländer,* no. 621.

115. *Smith Coll.,* no. 356, in blue, resting in a pointed amphora of the 1st C.

116. Kisa, *Glas* I 140–141.

117. Similar: Harden, *Karanis,* nos. 851ff. For a general discussion cf. Kisa, *Glas* I 139; Harden, *Karanis,* 282–283 (pointing out the difficulty of dating such minor objects). Cf. also T. E. Haevernick, *Die Glasarmringe und Ringperlen der Mittel und Spätlatènezeit auf dem europäischen Festland* (Bonn 1960).

(*BASOR* 166, 5ff.). The other bracelet, **233**, twisted to form a rope, was unstratified; a date within the Roman period, however, is likely.

219 Rod section, spirally twisted. Aquamarine.
P.H. 4.0, diam. 0.6.
BE 1964 vaulted chamber under MC ca. E33.50-E35/N57-N60 *91.97-91.27.

220 Upper portion of spirally twisted rod, flattened at top. Highly corroded.
P.H. 2.8, Th. 0.6.
HoB 1963 E0-E10/S120.5 to *101.40.

221 Upper portion of rod with collar at top. Greenish.
P.H. 2.7, diam. 0.7.
HoB grave 59.k E10/S60 *97.70.

222 G60.19:2724. Section of rod. Greenish?
P.H. 2.7.
HoB E10/S105 fill to *101.00.

223 Rod section spirally twisted to upper 1. Clear?
P.H. 4.8.
UT 1959 E50/S235 wall trench to ca. *124.10.

224 Front portion of pointed pin. Dark glass.
P.L. 3.3, diam. 0.3.
PN 1967 W275-W277/S329.5-S333.5 to *87.40.

225 Section of pin. Blue.
P.L. 6.5, diam. 0.5.
HoB 1961 E5/S95 *99.70-99.00.

226 Bulb-shaped pinhead, perforated from bottom. Aquamarine.
H. 1.2.
PN 1964 W250-W253/S347-S350 *87.60-87.00.

227 Tube section. Greenish.
BE 1964 vaulted chamber under MC ca. E33.50-E35/N57-N60 *90.93.

228 Tube section. Aquamarine.
L. 6.7, diam. 0.8.
MTE 1964 E65-E70/S155-S160 *109.00-108.50.

229 Spherical pendant (or bead), no perforation. Clear; brilliant iridescence; fractured at one end.
Diam. 1.4.
MTE 1964 E59-E65/S152-S155 to *111.40.

230 G63.9:5255. Tear-shaped with loop at top: probably a dripping (from manufacture) used as pendant. Greenish.
H. 1.7.
PN grave 63.1, W221/S347 *88.70.

231 J64.1:6108. Ring, flattened at one side. Greenish(?); black scum, corroded.
Diam. 1.7.
MTE E46-E50/S125-S130 *101.00-100.60.

232 G61.7:3181. Curved section of bracelet, ext. tooled to form notches. Black(?); rough surface.
P.L. 3.0.
HoB W5-W10/S90 to *102.00.

233 *Pl. 10*. G73.8:8267. Section of bracelet, flat at one end, twisted to form a rope. Olive green.
Est. diam. 8.0, Th. 0.8.
PN/E between graves 73.4 and 73.5, ca. W209.5-W215/S360-S364 *90.60.

III EARLY BYZANTINE GLASS

The majority of glass finds from Sardis is of Early Byzantine date, i.e. from ca. A.D. 400 to 616. Almost all the glass comes from occupation levels, not from graves and practically all the vessels have been broken into very small fragments. A very rough estimate of the number of fragments excavated between 1958 and 1978 seen by this writer totals about 7500 sherds. This sum, in turn, might represent about 4000 vessels if one assumes that approximately two fragments were part of each vessel. In addition, the remains of well over 1000 window panes should be added to this total. It should be noted that such an experienced excavator as D. G. Mitten estimates the ratio of pottery to glass to be about 15:1. Since the number of positively identified pre-Roman and Roman finds, including the bottles bought from a villager, amounts to little over 200 items, by far the majority of the glasses excavated at Sardis is of Early Byzantine date.

Selection Criteria

Practically all material with inventory numbers as well as almost all single items labeled by the excavator are included in the catalogue and the statistical lists.

The other material—exceeding by far in sheer quantity and weight the glass mentioned above—was investigated with the purpose of selecting identifiable and/or significant sherds (for example, well-preserved handles, feet, rim sections, fragments with thread or pattern molded decoration, etc.). A selection of these objects is included in the catalogue and in the statistics.

After this selection was made the boxes were sifted through in an attempt to identify sherds whose original vessel shape could be reconstructed. All of the fragments that could be identified were counted and entered in the statistics.

Finally, the contents of the boxes and cartons were duly noted and their weight estimated in an attempt to approximate the number of sherds contained in each. Again the totals were added to the statistics. The resulting grand totals (about 7500 fragments) can be found in the statistical table.

The Types

Save for a number of bottles, **476–478**, **488**, **509**, **560–565**, one beaker, **472**, and a salver, **374**, only a very few pieces could be reconstructed to show a continuous profile. Practically all of the finds consist of small fragments with dimensions only occasionally exceeding 3 cm. In some cases it was virtually impossible to say whether sherds identical or very similar in appearance were originally part of a particular vessel form or whether they represent different—though perhaps related—types. This seems to be particularly true of items such as the base rings of vessels (cf. **401–444**); many were excavated, but not one was large enough to reconstruct a complete vessel.

In computing and cataloguing the glass finds from Sardis, the margin of error, therefore, appears to be rather high. However, while the fact remains that inaccuracies in the identification of the material might give the conclusions presented below a slightly speculative twist, the damage does not seem to be too serious; we hope that the majority of the finds has been properly identified. Be that as it may, a survey of all the recorded sherds datable to the Early Byzantine period shows that over forty different vessel types are represented. This number does not include variants of goblet stems or base constructions that could not be linked to specific vessel forms.

Fabrics

Most of the Early Byzantine glass from Sardis seems to have been made in six fabrics of which four were quite popular and two relatively rare. An additional number of fragments have tints that seem to be variants of the main fabrics.

The most common, *fabric 1,* is light aquamarine; it is closely related to the typically Romano-Syrian blue green glass of middle and late Roman Imperial times, found most often in Syria. Next in frequency is a pale green glass, *fabric 2,* often tending towards the pale olive-green, and a "bottle" olive green glass, *fabric 3. Fabric 3* is darker than fabrics 1 and 2 and has a certain resemblance to ordinary eighteenth and nineteenth century olive green bottles made in England or Holland. *Fabric 4,* of the same intensity as fabric 3, apparently occurs less frequently; it has a "bottle" green color, i.e. a brighter green than fabric 3, that is also found in ordinary bottles of later times. *Fabric 5,* very rare indeed, is almost clear with a very slight green or yellow tint. About half a dozen vessel fragments are made of light blue glass, *fabric 6.*

Weathering

Almost every sherd shows some kind of weathering. A decay common on Early Byzantine glass at Sardis shows a dulled, frosted surface. This layer, in turn, may develop into an iridescent scum which flakes off easily. Occasionally the scum becomes much thicker and has a light ochre, enamel-like appearance.[1] A large number of fragments show the effects of a conflagration: the surface is covered with a dark, dirty layer that can only be the result of intense heat and contact with burning material. We have named this kind of decay "dirty fire scum" which, in most cases, covers a brilliant silver or peacock iridescence.

Manufacture at Sardis

All glass from Early Byzantine levels at Sardis displays a remarkable homogeneity that is generally found only at sites that either had manufacturing facilities of their own, or revealed hoards of glass imported from one particular region or factory. Lamps, goblets, and bottles as well as groups such as ring bases, **401–444,** or threaded bottle necks, **607–621,** are, in most cases, recorded in many hundreds of fragments which are all practically identical save for minor variations. Each piece is either made of a basic fabric, fabrics 1-4, or shows only slight deviations from the basic color or tint, the result of workshop methods typical for larger glass houses.

There seems to be no readily recognizable change in style within each vessel type or fabric over a period of ca. 200 years (assuming that the finds are more or less evenly spread over the entire period), a fact that speaks for a rather conservative tradition in a region situated on the periphery of the mainstream of glass development. It must be kept in mind, however, that development, in general, does not seem to have been very rapid. Since cullet, a few wasters and the remains of crucibles for the melting of glass were excavated in Early Byzantine levels (cf. **712–737**), there is no doubt that one or more workshops or factories were functioning in this city. These glass houses appear to have limited their production to the making of ordinary, undecorated or slightly decorated ware for home consumption while costlier vessels, bearing engraved, cut, or rich thread decoration, are totally lacking. Drinking and pouring vessels, lamps, bowls, and salvers belong to the regular output, accompanied by large amounts of window panes and tesserae for the decoration of interior walls of buildings (for example the Synagogue). While most of the production did not attain very high standards, some vessel types, for instance a variant of the goblet, **300–322,** were very carefully fashioned.

The decorative techniques employed for the glass appear to have been limited to simple spiral threading, **607–632,** and pattern molding in rib molds, **643–656.** Only a few examples of millefiori glass, **657–666,** if made locally, indicate experimentation with more complicated techniques.

The Significance of Sardis Finds

The importance of the Early Byzantine finds from Sardis lies in the fact that a large amount of well dated glass covering a relatively short period (ca. 400–616) was excavated under careful archaeological supervision. Because our knowledge of the history of glass

1. For the method of dating weathered glass, including samples from Sardis, cf. R. H. Brill, H. P. Hood, "A New Method for Dating Ancient Glass," *Nature* 189:4758 (Jan. 7, 1961) 12–14; R. H. Brill, "The Record of Time in Weathered Glass," *Archaeology* 14 (1961) 18–22.

and development of forms within this span of time—
in the Near East and under Byzantine rule—is still
relatively sketchy the finds presented in this volume
may help to give the reader a clearer view of the production of glass in Asia Minor and of the close stylistic
links it had with Syria and even northwest Italy. In
discussing each form in the catalogue, reference will
be made to parallel examples from other sites. The
following brief survey of similar glass finds may help
to orient the reader within the general framework of
post-Roman glass in the Near East.

The history of glass manufacture in the Near East
in late and post-Roman times has recently been described by J. Philippe and D. B. Harden in concise
surveys accompanied by bibliographical material.[2]
Glass of this period from Palestine was the subject of
D. Barag's dissertation.[3] From what is known of the
typology and modes of decoration in the latter part of
the fourth century, immediately preceding the Early
Byzantine period in Sardis,[4] the variety of shapes and
the decorative devices were still quite abundant. Both
experienced a regression towards the turn of the
fourth to the fifth century. Luxury ware does not
seem to have been in demand after the year 400 if
one excepts the richly cut glass produced for the Sassanian court. The embellishment of glass through
cutting, engraving, painting, mold blowing, and generous use of thread decoration came to a seemingly
abrupt end. Vessel shapes tended to become elongated. Lamps and stemmed drinking vessels became
more popular while receptacles such as beakers and
bowls lost their former relatively light and almost elegant forms. The well-known blue green Syrian glass
continued to be made in the centuries after the collapse of the Western Empire and Syrian vessel forms,
though reduced in number, served as prototypes for
the glass manufacture in the Near East after A.D.
400.[5]

For the convenience of the reader a number of references to excavation reports of sites having material directly related to the glass from Sardis are
listed in Table 3. References to specific vessel types
will be found in the introductory notes to the various
sections.

Some of the contemporary Sassanian finds from
Ctesiphon are no doubt closely related to a number of
vessel forms also found in Sardis.[6] Further west, archaeological sites have also revealed glass that in
many cases bears close resemblance to the Sardis Early
Byzantine ware. Apart from the finds from Corinth[7]
and Sucidava[8] the material recently excavated in Italy
and Southeast Europe should particularly be mentioned, and is presented in Table 4.

At most of these archaeological sites, many of
which were excavated after World War II, vessel
forms were found that are common at Sardis: lamps
of various types, stemmed goblets, long-necked bottles with globular bodies and concave bases, vessels
with ring bases, bottles with spiral thread around
the upper neck, occasionally pattern-molded ribbed
ware, and, of course, window glass and tesserae.

The Sardis finds thus confirm several suppositions
about glass production along the Eastern Mediterranean and in the Near East (and on the northwestern
coast of Italy) after the decline of Western Roman
power and before the advent of Islamic rule. The
misnamed "dark" period in the Near East saw much
more activity in glass manufacture than has been supposed. Even a provincial city such as Sardis had facilities to produce large quantities of admittedly ordinary ware for home consumption and possibly for
export. Glass made about the middle of the first millennium A.D. was part of an international style that
transcended national and geographical boundaries.
Influenced particularly by the Syrian-Palestinian con-
ventions of forms, fabrics, and modes of decoration
popular about A.D. 400, ordinary glass found (and
made) in Palestine, Asia Minor, Greece, and Italy is
close in date and appearance. The corpus of dated
finds—especially those from Sardis—now makes it
possible to describe accurately and in great detail the
situation from the fourth to the seventh century,
bridging the gap between middle and late Roman Imperial[9] and Islamic glass.[10]

Findspots

Areas where particularly rich Early Byzantine glass
finds came to light include the Byzantine Shops, RT
(south of shops E 12 and 13), portions of B (Pa, BE-A,
BE-B) and Syn, especially Fc. The vessel shapes listed

2. Philippe, passim; D. B. Harden, "Ancient Glass, II: Roman,"
ArchJ 126 (1969) 44–77; idem, *ArchJ*, 78ff.

3. Barag, "Glass Vessels"; cf. also A. von Saldern, "Ein gläserner
Schlangenkorb in Hamburg," *Festschrift für Peter Wilhelm Meister*
(Hamburg 1975) 56–61.

4. Cf. esp. the finds from Jeleme: Goldstein, passim.

5. Philippe, passim.

6. Puttrich-Reignard, passim.

7. J. Wiseman, "Excavations at Corinth, the Gymnasium Area,
1967–1968," *Hesperia* 38 (1969) 105–106.

8. D. Tudor, "Sucidava: Une cité daco-romaine et byzantine en
Dacie," *Collection Latomus* 80 (1965) 88, fig. 31.

9. Harden, *Karanis;* Clairmont, passim; Barag, "Glass Vessels";
Goldstein, passim.

10. Lamm, *Mittelalterl. Gläser.*

Table 3. Glass of about the same date and very similar in style to Early Byzantine glass from Sardis.

Site	Date of glass concentrations	References
Jerash	5th–6th C.	Crowfoot-Harden; Baur-Kraeling
Shavei Zion	5th–6th C.	Barag, *Shavei Zion*
Auja Hafir	Ca. 5th–7th C.	Harden, *Nessana*
Samaria	Ca. late 4th–6th C.	Crowfoot, *Samaria*
Mt. Nebo	Ca. 5th C.	Saller, *Moses*
Bethany	Ca. 5th–6th C.	Saller, *Bethany*
Khirbet el-Kerak	Early 5th–early 7th C.	Delougaz-Haines
Nikertai	5th–6th C.	Canivet
Hira	Perhaps 5th–7th C.	Unpublished; finds are in the Ashmolean Museum, Oxford.
Apollonia (near Cyrene)	Early Byzantine	University of Michigan excavation (lamps)
Beth Shan, Tell el-Hosn	4th–7th C.	FitzGerald
el-Bassa	4th–7th C.	J.H. Iliffe, "A Tomb at El Bassa of ca. A.D. 396," *QDAP* 3 (1933) 89
Mt. Olive	Ca. late 4th C. and later	Bagatti-Milik, 141–158.
Palmyra	Early Byzantine	Cf. Canivet, 66
Netiv Ha-Lamed He	Mid 5th–early 7th C.	Barag, "Netiv"
Kharjih	6th C.	Harden, *ArchJ*, 79, fig. 1:I, J
el-Jish	Ca. 5th–7th C.	N. Makhouly, "Rock-Cut Tombs at El Jish," *QDAP* 8 (1938) 45–50, pls. 32–33.
Beit Ras, Taluzza, Ajlun	Late 4th–early 5th (?) C.	D.B. Harden, "Some Tomb Groups of Late Roman Date in the Amman Museum," *Annales du 3ᵉ Congrès des "Journées Internationales du Verre," Damas, 1964* (Liège [1965]) 48–55; idem, *ArchJ*, 79, fig. 1.
Apamea (Syria)	Early Byzantine	Information: Renate Pirling. Unpublished (goblets, lamps).
Tell Hesbân (Heshbon)	Late Roman/Early Byzantine	S.M. Goldstein, "Glass Fragments from Tell Hesbân, A Preliminary Report," *Andrews University Seminary Studies* 14 (1976) 127–132.
Mezad Tamar	Late Roman/Early Byzantine 4th–5th C.	E. Erdmann, "Die Glasfunde von Mezad Tamar (Kasr Gehainije) in Israel," *Saalburg Jahrbuch* 34 (1977) 98–146.
Aphrodisias	Early Byzantine	Information kindly provided by Kenan T. Erim. Unpublished (goblets and cups).
Debeira West	7th–8th C.	D.B. Harden in P.L. and M. Shinnie, *Debeira West* (Warminster n.d.) 83–89.

in Table 5 are almost never represented by complete examples but by fragments. Only those shapes that occur frequently are listed. For the identification of these areas and for basic data on them, cf. *Sardis* R1 (1975) 13–16; Sardis M1 (1971) 2–5.

LAMPS

Lamps appear to have been more numerous than most other forms of Early Byzantine glass. This is surprising but might be explained by the fact that lamps

Table 4. Early Byzantine glass from Italy and Southeast Europe.

Site	Date of glass concentrations	References
Rome, Aventine Hill	Late 4th C.	C. Isings, "Some Late Roman Glass Fragments from Rome," *Bruxelles Congrès,* 262.2–3.
Invillino	5th–7th C.	Fingerlein et al.
Castelseprio	Post-Roman, Justinian	Leciejewicz et al.; S. Kurnatowski et al., "Gli scavi a Castelseprio nel 1963," *Rassegna Gallaratese di Storia e d'Arte* 27 (1968) 61–78.
Torcello	7th–8th C.	L. Leciejewicz, E. Tabaczynska, S. Tabaczynski, "Ricerche archeologiche nell'area della cattedrale di Torcello nel 1961," *Bollettino dell' Istituto di Storia della Società e dello Stato Veneziano* 3 (1961) 28–47; idem, ". . . Torcello nel 1962 . . . ," ibid. 5–6 (1963–64) 3–14; A. Gasparetto, "A proposito dell'officina vetraria Torcellana," *JGS* 9 (1967) 50–75.
Kastel Iatrus near Krivina (Bulgaria)	Late 5th–6th C.	Information: Gudrun Gomolka. Unpublished (goblets).
Salona	4th–5th C.	S. H. Auth, "Roman Glass" in C. W. Clairmont et al., *Excavations at Salona, Yugoslavia (1969–1972)* (Park Ridge, N.J. 1975) 166, no. 112, pls. 31, 33.

have certain more easily identifiable characteristics (handles, pointed base, etc.) and, therefore, figure more prominently in the statistics. The sherds of other vessels, having less pronounced features (bottles, beakers, etc.) cannot be connected with specific forms and for this reason have to be left out of the catalogue and the statistics.

At least five basic varieties of lamps were used in Sardis during the fifth to early seventh century. The types they represent have been described in Crowfoot-Harden, which to a great extent is based on the material found at Karanis (fourth to fifth century) and Jerash (fifth to eighth century).[11] Curiously, the conical beaker-type, frequent at other Near Eastern sites and often decorated with applied prunts and wheel-cut grooves, has as yet not been found at Sardis.[12] This fact, however, corresponds with the find pattern in other categories of Early Byzantine and, to a lesser extent, of Roman glass at this site. The primary objective of the workshops in this city was to supply a steady flow of ordinary glass products: window panes, tesserae for mosaics, lamps and a very limited number of vessels of more unusual shapes.

The Early Byzantine lamps are usually found with the remains of goblets, bowls, bottles, and window panes. Four or perhaps five forms appear to have been in use; the first two forms in our listing seem to be variations of one type.

11. Crowfoot-Harden, 196ff. Harden, *ArchJ,* 81. For conical 4th–5th C. lamps see also E. B. Dusenbery, "Ancient Glass in the Collections of Wheaton College," *JGS* 13 (1971) 23, nos. 36–37. For a survey of Islamic lamps see Lamm, *Samarra,* 30–35.

12. Isings, no. 106. Recently a series of conical vessels approximately contemporary with the Sardis material and often covered with geometric cut decoration has come to light, particularly in Iran; they may have been used as lamps as well as beakers, A. von Saldern, "Achaemenid and Sassanian Cut Glass," *Ars Orientalis* 5 (1963) 8, fig. 3. (Since this paper was published, a large amount of additional material has appeared on the market which I plan to treat in a forthcoming paper.) See also 9th C. conical and trumpet-shaped beakers found in Birka: Arbmann, pls. 189–192. For a conical bowl lamp cf. also M. C. Ross, "A Tenth Century Byzantine Glass Lamp," *Archaeology* 10 (1957) 59–60.

Table 5. Early Byzantine glass: findspot chart.

Sector	Early Byzantine glass types	Catalogue numbers
AcT	Base rings: folded	462
	Bottles	507, 555
	Bottles: cylindrical	652
	Folded feet of vessels	399
	Lamps	269
	Rims of vessels: folded	599, 600, 601
B-W	Bottles	496, 542, 559
	Threaded necks of vessels	614
	Vessels: with concave base	647
	Windows	689, 690
BE-A	Bottles: small	562
	Goblets	372, 373
	Lamps	283
	Miscellaneous stems	387
	Tesserae	704
	Windows	682, 693
BE-B	Base rings	416–419, 441
	Bottles	524
	Bottles: small	570
	Cullet	716, 722
	Goblets (over 100)	317, 328, 363, 441
	Lamps	267, 286
	Large dishes	465
	Millefiori	659
	Rims of vessels: plain	573, 586, 590
	Tesserae	706, 707
	Threaded necks of vessels	619
	Vessels: wide-necked with thread decoration	625
BE-C	Bottles	508
	Cullet	715, 725
	Lamps	234, 265
	Threaded necks of vessels	621
BE-E	Bottles	479, 480
	Bottles: globular with wide neck	653
BE-H	Bottles	482, 489, 494, 510, 511, 534
	Cullet	712
	Lamps	282
	Tesserae	703, 705
	Threaded necks of vessels	613, 616
	Vessels: wide-necked	623
BE-N	Base rings	443
	Beakers	475
	Lamps	258
	Miscellaneous stems	388
	Tesserae	701
	Vessels: with inlaid thread decoration	640
	Salvers	380

(Continued)

Table 5. Early Byzantine glass: findspot chart. (*Continued*)

Sector	Early Byzantine glass types	Catalogue numbers
BE-S	Bottles	585
	Cullet	713
	Goblets	346
	Lamps	262
	Threaded necks of vessels	615
	Vessels	656
BS E 1	Goblets	345
	Lamps	250, 280, 291
	Salver	378, 379
	Windows	683, 684
	Vessels: wide-necked with thread decoration	628
BS E 4	Base rings	401
	Cullet	721
	Goblets	369
	Lamps	239, 266
	Windows	685
BS E 5	Bottles	516
	Cullet	717
	Rims of vessels: folded	594
BS E 6	Windows (and lead strips for holding panes in place)	686
BS E 7	Windows	687
BS E 12	Bottles	484, 488, 528
	Goblets	368
	Vessels: wide-necked with thread decoration	632
	Windows (over 350 window panes, fabric 1 and variants of fabrics 2 and 3; particularly bubbly)	680, 688
BS E 13	The greatest amount of, and most varied, finds including almost all Sardis vessel forms; 1962 finds comprise ca. 4000 fragments: 9/10 vessels, 1/10 windows; 1/3 fabric 1, 1/3 fabric 2, 1/3 fabrics 3 and 4	
	Base rings: folded	454
	Bottles	495, 499, 502, 519–523, 525, 538, 549, 557, 561
	Goblets (over 150)	300, 323, 347, 348, 367, 371
	Lamps	242, 277
	Pattern-molded ware	643
	Rims of vessels:	
	folded	595–597
	plain	576, 577
	Rods	677
	Salvers	377
	Vessels: wide-necked with thread decoration	629
BS E 14	Lamps	240, 241
	Stamps	668, 669, 670

(*Continued*)

Table 5. Early Byzantine glass: findspot chart. (*Continued*)

Sector	Early Byzantine glass types	Catalogue numbers
BS E 15	Base rings: folded	456
	Beakers	474
	Bottles	558
	Bowls: high	395
	Goblets	340
BS E 16	Bottles	485, 503
	Lamps	263
	Rims of vessels: folded	603
	Rods	676
	Salver feet	375
	Vessels: wide-necked with thread decoration	622
BS E 17	Cullet	726
	Goblets	302, 303, 325
	Lamps	238, 264, 281, 297
	Large dishes	466
	Rims of vessels:	
	folded	602
	plain	584
BS E 18	Millefiori	661
	Tesserae	711
BS E 19	Base rings	444
	Bottles	486
	Lamps	295
	Rims of vessels: with heavy threads	633
BS W 1–2	Bottles: small	560
	Goblets	310, 311
	Lamps	260
	Pattern-molded ware	654
	Rims of vessels: folded	593
	Windows	681
BS W 3	Base rings	402–404
	Threaded necks of vessels	608
BS W 7	Flat bases	445
BS W 8–9	Bottles: small	568, 569
BS W 10	Base rings	405
BS W 13	Base rings: folded	458
	Bottles	497
	Cullet	734
	Goblets	315, 316
	Lamps	287
	Rims of vessels: plain	587
BS W 14	Base rings: folded	455
	Bracelet	679
	Goblets	318

(*Continued*)

Table 5. Early Byzantine glass: findspot chart. (*Continued*)

Sector	Early Byzantine glass types	Catalogue numbers
BSH	Vessels	655
	Vessels: with inlaid thread decoration	642
HoB	Base rings	420–437, 442, 470
	Base rings: folded	459
	Beakers	472
	Bottles	530, 535, 536, 543, 545
	Bowls: footed	383
	Cullet	719, 728
	Goblets	319–321, 331–333, 349–351, 366
	Lamps	247, 254, 272, 276, 288, 292
	Large dishes	467–471
	Millefiori	663
	Miscellaneous stems	390
	Rims of vessels:	
	folded	604
	with heavy threads	634, 635
	plain	572, 575, 578, 579, 582
	Salvers	374
	Tesserae	709
	Threaded necks of vessels	607, 612
LNH	Bottles	506, 567
	Lamps	290
	Rims of vessels: folded	606
	Tesserae	700
MC	Goblets	357, 358
	Lamps	259
	Vessels (many fragments of unidentified form)	
MTE	Base rings: folded	460, 461
	Bottles	527, 531, 551
	Bowls: shallow with waisted rims	391, 393
	Cullet	723, 724, 727
	Lamps	273, 275, 289
	Rims of vessels:	
	with heavy threads	636
	plain	591
Pa-S	Bottles	491, 539, 547, 548, 571
	Bottles: curse	571
	Lamps	278
	Windows	691, 692
Pa-W	Bottles	532, 533
	Lamps	298
	Miscellaneous stems	384, 385
	Salver feet	376
	Stamps	672
	Threaded necks of vessels	609, 610, 611, 617
PN	Base rings	400, 440
	Bottles	481, 487, 517, 518, 540, 544, 552, 553

(*Continued*)

Table 5. Early Byzantine glass: findspot chart. (*Continued*)

Sector	Early Byzantine glass types	Catalogue numbers
	Cullet	718, 720
	Flat bases	447
	Goblets	334–336, 352–354, 364, 365
	Lamps	237, 257
	Miscellaneous stems	389
	Pattern-molded ware	646
	Threaded necks of vessels	620
	Vessels:	
	with inlaid thread decoration	641
	wide-necked	624
	Windows	697, 698
RTE	Base rings: folded	457
	Bottles	537, 564
	Miscellaneous stems	386
	Rims of vessels: plain	574
RTW	Rims of vessels: plain	580
	Stamps	671
Syn	Base rings	406–409, 411, 412
	Base rings: folded	449–451
	Beakers	473
	Cullet	736
	Folded feet of vessels	397
	Goblets	301, 305, 339, 341, 344, 356
	Lamps	248, 249, 251–253, 270, 294
	Millefiori	657, 658, 664–666
	Rims of vessels: plain	583
Syn Fc	Base rings: folded	452, 453
	Bottles	492, 493, 529
	Bracelets	678
	Cullet	729, 732
	Goblets	324, 355, 410
	Lamps	243, 244, 304
	Rims of vessels: plain	588
	Rods	675
	Stamp	673
Syn MH	Bottles	556
	Bowls: shallow with waisted rim	392
	Cullet	731, 735
	Tesserae	702
	Vessels: wide-necked with thread decoration	627
Syn porch	Bottles	645, 649
	Goblets	413
	Rims of vessels: folded	605

Type 1 Shallow bowl with handles
Type 2 Deep bowl with handles
    Type 2a: variant with rim bent inward
    Type 2b: variant with rim bent outward
Type 3 Lamp with cup-shaped bowl and stem- or
    thorn-like lower part
Type 4 Straight-sided or bell-shaped beaker with
    knob base
Type 5 Bell-shaped or funnel bowl

The shallow bowl lamp (type 1) appears to have been the least common form at Sardis. It is extremely fortunate that the entire height of the wall section of one lamp, together with a handle, has been preserved, thus making it possible to present an accurate reconstruction of the vessel.

There is no way of estimating how frequent the deep bowl lamp (type 2) was at Sardis. Though some rim fragments and a great number of small handles were found, one cannot be certain how many of these handles were actually part of lamps and, if so, how many of them belonged to type 1 or type 2 or to a beaker-shaped lamp. Incidentally, the incorporation of handles in a statistical survey tends to distort somewhat our view of the Early Byzantine glass at Sardis. Vessel sections with a handle have a better chance of survival and of recognition as such than the less sturdy wall sections. Small handles of a particular shape seem, in our view, to have belonged to lamps—and were therefore recorded in the catalogue as well as in the statistics—while curved wall sections from vessels of unidentifiable shape had to be left out of the catalogue.

The remains of only two bowl lamps of the shallow variety (type 1) are recorded; in contrast, sections of over a dozen deep bowl lamps of type 2 could be identified. In addition at least 150 single handles were registered, most of which we believe were part of bowls of type 2.

Type 3, a small bowl with a tube- or thorn-like extension, was probably as, or even more, frequent than type 2. Well over 100 sections of these lamps are recorded, and certainly many more are left unidentified in the piles of fragments.

The beaker-shaped lamp type 4 is represented by about two to three dozen vessel fragments, which seems to indicate that this form was less popular than types 2 and 3.

As pointed out above, the importance of statistical data might lie not so much in the fact that a specific number of objects of one type was found at the site but rather in the relative frequency of each type. Thus the ratio of type 1 to type 2 appears to have been at least 1:10, that of type 2 to type 3, 1:1 or 1:2, that of type 3 to type 4, 2:1. The total number of lamps in relation to goblets seems to have been about 1:2.

A roughly bell-shaped bowl pointed at bottom may also represent a lamp, here listed as type 5. The base fracture could, however, be an indication that it once had a foot—a fact that might speak against its use as a lamp.

Type 1

The shallow bowl type with flat base and slanted walls is closely related to the deep bowl lamp, type 2.[13] Of the two examples found at Sardis, one is of greenish glass (probably fabric 2), **234**, and the other of aquamarine glass (fabric 1), **235**. Both have estimated diameters of ca. 11 cm., and the height of their rims is about 5 cm. The lamps have been found in Early Byzantine contexts; **234** comes from BE-C, and can be associated with the period after the remodeling in the fifth century. As coins of Constans II (A.D. 641–668) were also found nearby, one might be inclined to put the date of the lamp in the latter part of the occupation, i.e. the sixth to early seventh century (*BASOR* 187, 19–20).[14] The other piece, **235**, was found nearby in BS E 1–2 and is certainly of the same period.

234 *Pls. 11, 23.* G66.11:7263. One third with concave bottom, flaring wall and 3 handles attached at bottom and rim; the rim is bent inwards where handles join edge. Fabric 2.
P.H. 5, diam. base 9, est. diam. rim 11.5.
BE-C E18-E19/N11.7-N12.5, floor to *96.15.

235 *Pl. 23.* G68.2:7686. Two thirds of lower portion (broken in two), flat bottom, flaring wall, 3 handles. Fabric 1.
P.H. 3.5; diam. at fracture 8.5; est. diam. of base 6, of rim ca. 11; Th. 0.2.
BS E 1-2, E12-E13.50/S0-S4 *97.12-96.80.

236 G73.1B,C:8233. Two fragments with out-turned rims. Greenish.
Max. dim. 6.9; est. diam. ca. 11.0.
PN/EA W208/S364, in bedding of earlier church floor, *89.56. Found with **246** and **400**.

13. For close parallel examples from Jerash, datable to the 4th–5th C., see Baur-Kraeling," type H," 527ff., fig. 23:34, 36, 37.
14. For the coins, *Sardis* M1 (1971) 158, and the listing under BE-C on p. 149.

Type 2

This bowl lamp is more common than type 1. Sections of over a dozen lamps are catalogued here to which should be added numerous handles that certainly must have been part of identical or at least very similar vessels. Not a single complete example was found nor could a combination of vessel fragments be assembled that would show a profile of a complete vessel. We have to rely, therefore, on parallel pieces found elsewhere. The best known variety of the lamp has a fairly deep, generally straight-sided bowl tapering downwards, with three small handles attached to the rim. The rim may flare outward or, less frequently, be bent inward.[15] In analogy to type 1, bowl lamps of type 2 may have had flat or concave bases.[16] A few of those found at Sardis have a pattern-molded rib decoration identical to that of pattern-molded bottles (cf. 643–656).

The lamps are made of aquamarine (fabric 1) and greenish (fabric 2) glass. The preponderance of aquamarine handles seems to indicate preference for fabric 1. The few portions of rims that are reconstructible show that the diameters vary from 9 to 12 cm. while the total height seems to have measured about 8 to 10 cm.

Since we believe that most of the small handles found in the excavations belonged to lamps of type 2, they have been included in this chapter. Their height varies from 3.5 to 5 cm., with a median height of 4 cm.[17] At the end of this chapter a few handles are listed that are different from the others but might also have belonged to identical or similar lamps.

The largest preserved section of a lamp, 237, was found in the bath at PN, which has been provisionally dated to the fifth century (BASOR 166, 16–18; Sardis M4 [1976] 46). With it were found portions of Early Byzantine bottles, 487 and 518. This lamp has pattern-molded ribs running from lower right to upper left (cf. 643–656) and a reversed rib pattern is shown on 238. This piece was found in BS E 13 in a subfloor level which apparently predated the material generally coming from other Byzantine Shops (BASOR 174, 45–46), and might then be datable to the fifth century. With it came sherds of plain vessel rims of fabric 1 and window fragments, whereas the level above revealed remains of goblets (for example, 303).

Two other pattern-molded lamp sections, 239, were excavated in BS E 4, which, together with neighboring shops, was in use in the sixth and early seventh century (BASOR 191, 17). With them were found various aquamarine-colored vessel fragments, including the shoulder section of a bottle with conical neck (type 3, 489–499), window panes covered with dirty fire scum, and the section of a multiknobbed goblet(?) stem, 369.

More numerous are the bowl lamps without pattern-molded decoration; most of them come from the

15. For late 4th C. predecessors very close in form to the Sardis lamps see C. Isings," Some Late Roman Glass Fragments from Rome," *Bruxelles Congrès*, 262.1, fig. 1:1–2. Isings, no. 134, cites other 4th–5th C. Western examples. A number of bowl lamps with handles, some of which rest on feet similar to goblet feet found in Sardis, are of late Roman and Early Byzantine date: Barag, "Glass Vessels," pls. 23, 28, 29, 40. For Early Byzantine bowl lamps with three handles and generally with folded rims see esp. Crowfoot-Harden, 201, pl. 28:7, from Gezer, ca. 5th–6th C. (cf. here R. A. S. Macalister, *The Excavation of Gezer* [London 1912] I 362ff., fig. 189). Ibid., 205, pl. 30:40–41, from Jerash, ca. 4th–5th C. For lamps from Jerash see also Baur-Kraeling, 526, nos. 29ff., fig. 22 (no. 29, here *Pl. 23*); a similarly shaped bowl with foot (524, no. 17, fig. 20:17) may also have been a lamp. (Some of the material from Jerash is preserved in Oxford, Ashmolean Museum, one lamp having a diam. of 8 cm.). Other bowl lamps of identical or very similar form: Crowfoot, *Samaria*, 418–419, fig. 99:2–3, ca. 5th–6th C.; Auja Hafir, 5th–7th C. (Harden, *Nessana*, 84, nos. 47ff., pl. 20; some material in Oxford, Ashmolean Museum); Barag, *Shavei Zion*, 68, nos. 21–22, fig. 16 (here *Pl. 23*), 5th–6th C.; handles from lamps from Nikertai, 5th–6th C. (Canivet, 65, figs. 8–9); Mount Nebo, 5th–6th C. (Saller, *Moses*, 317; idem, *Bethany*, 330, 5th C. Bowl lamps from Cyprus (Cambridge, Fitzwilliam Museum, inv. no. GR 32.1888) and Volubilis (J. Boube," Volubilis: Une lampe en verre du IVe siècle," *Bulletin d'Archéologie Marocaine* 4 [1960] 508–512, pl. 20, fig. 11) are very close to the lamps just mentioned but seem to be very late Roman, i.e. very late 4th C. An undated example: G. Lehrer, "Three Fragments of Rare Glass Vessels from the Museum Collection," Museum Haaretz, *Bulletin* 14 (Dec. 1972) 133–135, fig. 8–10. Cf. also S. H. Auth in C. W. Clairmont et al., *Excavations at Salona, Yugoslavia (1969–1972)* (Park Ridge, N.J. 1975) 166, no. 112, pl. 31 (late Roman); E. Erdmann, "Die Glasfunde von Mezad Tamar (Kasr Gehainije) in Israel," *Saalburg Jahrbuch* 34 (1977) 100, 112–113, nos. 3 ff., pl. 1 (4th–5th C.); F. Fremersdorf, *Antikes, islamisches und mittelalterliches Glas . . . in den Vatikanischen Sammlungen Roms* (Vatican City 1975) pl. 61 (lamp in the Römisch-Germanisches Zentralmuseum, Mainz); R. Rosenthal and R. Siram, *Ancient Lamps in the Schloessinger Collection* (Jerusalem 1978) no. 678. A ribbed, mold-blown lamp in Newark: S. H. Auth, *Ancient Glass at the Newark Museum* (Newark 1976) no. 197. A different type of lamp (?) of the same period, i.e. about the 5th C., is represented by a mold-blown, ribbed vessel found recently at Samos; it consists of a high foot, a squat lower body and a high and conical neck, apparently with applied handles (information kindly provided by Mr. Stechert). For 9th C. examples with straight walls and high stems, see R. H. Pinder-Wilson and G. I. Scanlon, "Glass Finds from Fustat, 1964–71," *JGS* 15 (1973) 22–23, no. 14, figs. 20–21. Similarly shaped lamps with small handles attached to the rim come from Corinth, datable to the 11th and 12th C., Davidson, *Corinth* XII nos. 733–734, fig. 14.

16. However, some of the fragments found may have come from bowl lamps with folded rim and three handles, resting on feet similar to those of the goblets (cf. 300–373); see Barag, "Glass Vessels," pls. 28, 40.

17. Lamm, *Samarra*, nos. 129ff. has also identified single handles as belonging to lamps.

Synagogue area. One large section, **240**, was found in BS E 14 in the fill of a drain (*BASOR* 170, 50). Among the coins associated with this level were those of Constantius II (A.D. 346-361) and Maurice (A.D. 582–602; *Sardis* M1 [1971] no. 603). **240** most probably belongs to the later part of this time span—the late sixth to early seventh century. Nearby were found various lamp handles, **241**. Another rim section, **242**, comes from BS E 13; at this spot a number of vessel fragments were excavated in 1963 (the shop was excavated in 1962, *BASOR* 170, 50) which had formed parts of goblets, lamps of type 3, footed cups (cf. **401–444**) as well as plain and pattern-molded bottles (however, the glass excavated in this shop in 1963 also included Roman Imperial glass as, for example, a fragment of a vessel with cut grooves, **58**).

Two slight variants of bowl lamp type 2 are represented by a few rim fragments recovered in the same general area and datable also to the Early Byzantine period. One, type 2a, has a rim bent inward while the second variant, type 2b, has a waisted upper wall with the lip turned outward, probably similar to a lamp found at Jerash.[18]

Of the variant with rims bent inward (type 2a) **243** and **244** were found in the Synagogue forecourt and can probably be associated with the remodeling of the area in the fifth century (*BASOR* 191, 30–31). In this area were also discovered various handles and the remains of a number of goblets of different types (cf. **300–373**), including one with a knobbed stem (type 2).

An example of variant 2b, **246**, was found at PN in 1973. A second, from HoB, **247**, dates to the sixth or early seventh century (*BASOR* 157, 22ff.). A similar piece, **248**, though of a different fabric, comes from the "closet" (chamber d) near the apse of the Synagogue. It was found with **249**. Related to it is a fragment with out-bent rim, **250**, from BS E 1 which contained window glass, vessel fragments, a red-ware plate attributed to the fifth century and "more than 100 coins ranging from Arcadius and Honorius (A.D. 395–408) to Phocas (602–610)" (*BASOR* 191, 17; *Sardis* M1 [1971] see listing on p. 149). This fragment is probably of fifth to sixth century date.

Numerous handles bent at an angle, having an average height of 4 cm. and generally made of fabric 1, were found all over the Synagogue area. Form, size, and color link them closely to lamp type 2 (the few handles that are larger could conceivably have belonged to type 1). In the following, a fair sampling instead of a complete listing will be given; the criteria for the selection are state of preservation and a datable context.

Some of the handles were discovered in Fc and the area east of it in Early Byzantine levels (**251–253**); others were found—together with lamps of type 3 and tesserae—at RT (**255–256**) at levels datable to the sixth and early seventh centuries (*BASOR* 166, 40–44). Another, **258**, comes from BE-N above the floor level of ca. A.D. 400 where coins of Leo I (A.D. 475) and Justin I (A.D. 518–527) help to date this level to the latter part of the fifth and the sixth century (*BASOR* 187, 57). Close by, in the vaulted chamber of B below the level of MC, another handle was found, **259**, which may belong to the Early Byzantine period; other finds at this spot include a lamp section of type 3 and various other Early Byzantine vessel fragments. Among the latter are the base of a beaker (?) with slightly conical lower part, and a tube which may be of fourth century date. This level seems to have been disturbed as Early Byzantine and late Roman glass appear to be mixed.

More securely dated is another handle, **260**, from BS W 2 where goblet sections and a hoard of coins of Heraclius (A.D. 610–641) came to light (*BASOR* 154, 17–18; *Sardis* M1 [1971] 154, "Hoard FF"). Other handles which may belong to these lamps were found in Early Byzantine levels in the Syn area, RT, Building B, and a number of Byzantine Shops, including E 14 and E 16.

A few handles come from the area east of Fc where numerous glass fragments, particularly of goblets, were found in 1963–1965, **261**. They differ from the other handles just referred to in that they are mainly of greenish (fabric 2), bottle olive (fabric 3), and bottle green (fabric 4) glass. With them were found cup bases (cf. **401–444**) and other vessel fragments of Early Byzantine origin.

Two handles, one attached to a rim spreading outward and apparently having belonged to a bowl lamp of either type 1 or 2, were discovered in BE-S with a corrugated handle, **262**. Both BS E 16 and the BS E 17 (*BASOR* 174, 45) contained handles and many vessel fragments (**263** and **264**), among them bottle necks with spiral threads (**622**) and bases of cylindrical vials (cf. **549–559**); also from these shops are sections of pattern-molded ware, part of the rim of a salver, sherds from goblets, and window panes.

A three-stranded handle, **265**, found in BE-C with a vessel base (cf. **401–444**) was in a level associated with the sixth or early seventh century (*BASOR* 187, 19–20, the unit designated C'). Two larger handles were found. One, **266**, may have belonged to a lamp

---

18. Crowfoot-Harden, pl. 30:41.

of type 1 or a very large example of type 2; it comes from BS E 4 (*BASOR* 191, 17). The other finds at that location include a bottle neck with applied spiral thread (cf. **607–621**) and plain conical bottle necks (cf. **489–499**) which are all of Early Byzantine origin. The other handle, **267**, was found in BE-B (*BASOR* 187, 15ff., the unit designated B′) at a level probably datable to the early fifth century; window panes and vessel fragments were found nearby. An object looking like one half of a massive handle, roughly prong-shaped, **268**, comes from the square room east of Fc (*BASOR* 174, 47) which contained various fragments and other material datable through coins ranging from Constantine I to Justinian. The piece could have formed part of an exceptionally large lamp.

The handle or rim section of a vessel, **269**, has a wavy, openwork thread attached. It may have been part of a bowl or lamp of a type unknown to us. A vessel with bulging bottom and wide, spreading neck which has handles with applied threads attached in a similar way[19]—a forerunner of enameled mosque lamps of the late thirteenth century—comes to mind.

A small group of handles tooled to form two distinct ribs with a "valley" between them are no doubt Early Byzantine. One was found in the center of the Synagogue, **270**, while two others, **271**, come from a spot close to the colonnade east of Syn (*BASOR* 174, 46). **272** and **273**, are not stratified; **273** could be Roman Imperial as it was found at MTE.

237 *Pls. 11, 23.* G61.18:3770. Upper portion and most of rim with conical wall, out-folded rim and 3 handles; pattern-molded with ribs turning to upper l. Fabric 1, eggshell; white scum; mended.
P.H. 4.5, diam. 13, Th. to 0.1.
PN W265/S360 *88.45-88.33. Found with **487** and **518**.

238 *Pl. 23.* One fifth of upper portion, part of one handle preserved; pattern-molded with ribs turning to r. Fabric 1, eggshell; similar to **237**.
Est. diam. 12, max. diam. 6.
BS E 17, 1963 subfloor *96.40-95.80.

239 Two rim fragments, same as **237** and **238**. Pattern-molded with ribs turning to upper r.
Est. diam. 13.
BS E 4, 1967 E29-E30/S0-S4 *98.00-97.20.

240 *Pl. 11.* G62.8:4227. One third of upper portion and rim, conical wall, folded rim, one (of 3) handles preserved; well made. Yellow tint, eggshell; slight iridescence, brown scum.
P.H. 5, est. diam. 9.
BS E 14, E84/S1-S1.5 *95.00, fill in drain. Found with it

were a steelyard and a pilgrim flask (P62.49:4240) with Pantocrator and Virgin (?); both would be of the 5th C. A.D.

241 Group of fragments with handles, some fire-twisted. RTE 1962 E80-E82/S5.5-S7.0, fill outside BS E 14 *96.80.

242 One fifth of folded rim with flaring wall, one handle preserved. Yellow tint, eggshell; heavy white scum.
P.H. 5, est. diam. 12, max. diam. 6.5.
BS E 13, 1963 E75-E80/S0-S3.

243 *Pl. 23.* One tenth of upper part, conical wall with plain rim bent inwards; probably 3 handles; well made. Very similar to **244**. Fabric 4; black fire scum.
P.H. 4.3, est. diam. 18, max. dim. 5.8, Th. to 0.1.
Syn Fc 1965 E104-E110/N1.20-N4.00 *96.60-96.40.

244 One eighth of rim preserved. Similar to **243**. Fabric 3.
P.H. 3.5.
Syn Fc 1967 E102.50-E105/N15.80-N19.50 *96.50-96.30.

245 One third of upper portion of vessel with conical wall and rim bent inward. Fabric 3.
P.H. 7.7, est. diam. 13, Th. 1.
Chance find, 1964.

246 *Pls. 11, 23.* G73.1D,E:8233. One fragment with out-turned rim, one small attached handle preserved. A similar handle (*Pl. 11*) may come from the same vessel. Heavily weathered.
Max. dim. 4.7, est. diam. ca. 11.0.
PN/EA W208/S364, in bedding of earlier church floor, *89.56. Found with **236** and **400**. The fragments were found with a ring base (G73.1F.8233; cf. **401–444**) as well as with the very heavily weathered base of a beaker. The author doubts that the rim fragment and ring base belonged to the same vessel. It is conceivable that the original vessel was not a lamp.

247 *Pl. 11.* G59.32a:1636. Portion of rim section with wall slightly tapering to plain rim, one (of 3) handles preserved. Fabric 1.
H. handle 3.
HoB E30/S60-S70 *97.24.
AJA 66, 11, no. 18, pl. 10:18

248 G63.2A:5034. One tenth of bowl with convex body and wide rim bent outward, one (of 3) handles preserved. Amber (olive-green).
P.H. 4, est. diam. rim 8, H. handle 3.
Syn E33-E38/N1.20-N2.5, "closet," lower fill *97.30-97.00. Found with **249**.

19. Ibid., pl. 30:47.

Handles

Predominantly fabric 1, median H. 4.:

249 *Pl. 11*. G63.2B:5034. Two-stranded; green; black scum.
H. 4.5.
Syn E33-E38/N1.20-N2.5, "closet," lower fill *97.30-97.00. Found with **248**.

250 *Pl. 23*. With attached portion of wall that curves outward at lower handle. Fabric 1 or 2; dirty fire scum.
H. 5.
BS E 1, 1967 E7-E10/S0-S4 *97.30-97.20.

251 Syn 1963 E105/N1.10-N4.0, water channel *97.25.

252 Syn 1964 E125/S5-S6 *98.00.

253 Syn 1964 E118-E121/S6-S8 *97.50-97.00.

254 HoB 1964 W22-W22.50/S118-S119 to *97.00.

255 RT 1961 E16-E17/S9, on colonnade floor *96.50. Found with **285**.

256 RT 1961 E18-E19/N13-N14 *96.50-95.00. Found with clear mosaic tesserae.

257 PN 1964 W274-W276/S341-S345 *88.85-88.70.

258 Fabric 2.
BE-N 1966 E25-E26/N84-N90 to *96.60 floor.

259 Vaulted chamber under MC 1964 *90.93. Found with bases of lamps of type 3 and bases of other vessels, including beakers (?) and bowls.

260 G58.38:538. Fabric 2.
BS W 2, W7-W8/S0-S3 *95.80, below top of S wall.

261 Four handles; fabrics 2, 3, and 4.
E of Syn 1965 E120.5-E125/N11-N15 *96.50-96.25.

262 With small portion of vessel. Decolorized.
BE-S 1965 E26.75-E29/N27-N36.25 to *96.10.

263 Fabric 2.
BS E 16, 1965 E92-E96/S0-S2 *97.25-96.50. Found with **622**.

264 Two handles; fabrics 1 and 2.
BS E 17, 1963 E97-E101/S1-S4 *96.60-95.90.

265 Three-stranded.
BE-C 1966 E17-E20/N10.7-N12, "floor to 20 cm. above."

266 *Pl. 23*. Unusually large handle of the type represented by the smaller examples, **249–265**; part of vessel wall preserved. Fabric 1.
H. 8.3, diam. handle 1, Th. wall 0.1.
BS E 4, 1967 E34-E35.50/S0-S4 *97.20-96.00.

267 Portion of large handle. Fabric 2.
P.H. 5.5.
BE-B 1966 E28.80-E30.55/N4-N7 *97.00-96.40.

268 Fractured prong-like object, handle? Fabric 1.
P.H. 3, diam. 2.3.
E of Syn 1963 E117-E121/N2-N4 *97.75-96.75.

269 *Pl. 11*. G60.7:2633. Curved section of handle (or vessel rim?) with applied, pincered snake-like thread. Fabric 2.
P.H. 4.6.
AcT D-E/1-3 fill, *100.75.

270 Lower portion of two-ply handle. Fabric 1.
P.H. 5, W. 2.5.
Syn 1965 E72-E73.5/N14-N15 *95.70.

271 Portions of 2 handles.
P.H. 5, 4.5; W. ca. 1.5.
E of Syn 1963 E118/N0-S1 *97.00-96.75.

272 *Pl. 23*. Two-ply handle widening at lower end where it is attached to portion of vessel wall.
P.H. 6, W. at lower end 4, Th. vessel wall 0.1.
HoB E5-E10/S120 *101.30.

273 Two-ply handle with portion of vessel wall.
P.H. 2.8, Th. vessel wall 0.1.
MTE 1964 E64-E72/S155-S161 to *111.20.

Type 3

Lamps with cup-shaped upper portion and stem- or thorn-like extension at the bottom were probably the most popular form from late Roman Imperial times and throughout Byzantine, Islamic, and western medieval periods. In the Near East this shape is still in use today. Probably derived from an almost identical vessel form used as a funnel in early Roman Imperial times,[20] it seems to have reached universal popularity in the Early Byzantine period:[21] it was fre-

20. Isings, no. 74. Morin-Jean, 146–147, form no. 117. Cf. also Vessberg, *OpusArch*, 151–152 (undated). For undated examples from South Russia cf. F. Rademacher, *Die deutschen Glaser des Mittelalters* (Berlin 1933) pl. 18. For late 4th C. examples see Isings (supra n. 15) 262.2, fig. 1:4.
21. Crowfoot-Harden, 198ff., have named this form B 2; see esp. pl. 29:24–25 (here *Pl. 23*).

quent at Jerash during the fourth to fifth century[22] and is known from many other sites.[23] In similar or slightly modified form it appears in ninth century Samarra[24] and in San Saba[25] in Rome before the twelfth century. Lamps of the tenth to eleventh century from Fustat, having a conical bowl and a long, thorn-like stem with terminal knob, represent another variant.[26] Pictorial sources in medieval Europe also show lamps of type 3.[27]

The lamps were either suspended from the ceiling by chains—most probably the exception—or set in hoops or metal (bronze) candelabra and polycandela, into which holes are cut to hold them. More elaborate candelabra include arms in the form of dolphins.[28]

A simple hoop with six cutouts and triple chain was found at Sardis in unit 13 of HoB, datable to the sixth or early seventh century.[29] Others, with openings for three,[30] four,[31] five,[32] six,[33] eight,[34] and even ten[35] lamps occur frequently. Most of these polycandela form a complicated, starlike network, often having Christian symbols and inscriptions;[36] a silver example bears a stamp of Justin II (A.D. 565–578).[37] We have a detailed description of the meaning of such candelabra in church services (and, we may add, also in the ritual of a synagogue) in the *Descriptio Sanctae Sophiae* of A.D. 563 by Paulus Silentiarios: suspended from the dome were chains with silver discs, ". . . these discs, pendent from their lofty courses, form a coronet above the heads of men. They have been pierced too by the weapon of the skillful workman in order that they may receive shafts of fire-wrought glass, and hold light on high for men at night."[38]

At Sardis not one single example of a lamp of type 3 was complete enough to show an entire cross section or profile. While the lower part—a thorn-like bottom with smooth underside from which the pontil rod was carefully removed—of many lamps is preserved, no section of the bowl-like upper part was discovered among the fragments that would show a continuous wall element up to the rim. There is no doubt, however, that a large number of unidentifiable wall and rim sections actually belonged to type 3 lamps.

The absence of a whole lamp makes it fairly difficult to give reliable information on the average complete height or the average height and diameter of the bowl portion. Nor do we know whether the bowl was straight, oval, spreading, or with a flaring rim. According to the curvature of the waisted sections of a number of Sardis lamps, and by analogy to lamps published by Crowfoot-Harden and others, we may assume that the most common form in Sardis was a thorn-like lower part gently flaring (trumpet-like) to an almost horizontal wall, followed by a sharp upward

bend, and terminating in a vertical, straight-sided bowl with plain, perhaps slightly flaring rim (see **274** and **275**).

The average height must have been about 8 to 10 cm., the diameter of the rim ca. 7 cm. The base diameter varies from 0.5 to 1.0 cm. (**284** with a base diameter of 1.5 cm., was certainly larger than the average

22. Ibid; Baur-Kraeling, "type E"(?), 523ff., fig. 17:14. Cf. the holdings in Oxford, Ashmolean Museum; the estimated height of these lamps is ca. 12 cm.
23. For example, fragments from Auja Hafir and Hira (Iraq) in Oxford, Ashmolean Museum; the lamps from Hira have a terminal knob. Cf. also Barag, *Shavei Zion*, 68–69, no. 25, fig. 16 (here *Pl. 23*) (5th–6th C.); Crowfoot, *Samaria*, 414ff., figs. 96–97 (4th–6th C.); Harden, *Nessana*, 84–85, nos. 51ff., pl. 20, (ca. 5th–7th C.); Saller, *Bethany*, 330, (5th C.); idem, *Moses*, 316–317, pl. 140; from Apamea, Syria (5th–6th C.; information kindly supplied by Renate Pirling). Many collections have lamps of this general type; cf. esp. the material in the Museum Haaretz, Tel Aviv. For Western examples cf. Fingerlein et al., fig. 13:12; Fremersdorf (supra n. 15) 95–96, nos. 875ff., pls. 60–61.
24. Lamm, *Samarra*, 33, nos. 145–147, fig. 26, pl. 4.
25. Crowfoot-Harden, 203–204; Lamm, *Samarra*, 78. For a complete example with an attachment for three chains, see M. E. Cannizzaro and I. C. Gavini, "Nuove scoperte avvenute nella chiesa di S. Saba, sul falso Aventino," *NSc* (1902) 273.
26. Pinder-Wilson and Scanlon (supra n. 15) 22, figs. 18–19.
27. Rademacher (supra n. 20) 79ff.
28. Ibid., 76.
29. *BASOR* 157, 24, fig. 12. Metal lamp holders from Sardis, most found in BS, will be published by J. C. Waldbaum in a forthcoming monograph in this series.
30. *JGS* 6 (1964) 158, no. 12.
31. Formerly Berlin; O. Wulf, *Altchristliche und mittelalterliche, byzantinische und italienische Bildwerke* (Berlin 1909) I no. 1006.
32. Ibid., no. 1004.
33. For example, ibid. nos. 1005, 1007; Walters Art Gallery, Baltimore, 54.2363 (*The Arts of Islam*, exhibition catalogue Hayward Gallery, London [London 1976] no. 182, Persia, 12th–13th C.). Galerie am Neumarkt, Zurich, auction no. 20 (Nov. 19, 1970).
34. Paris, Louvre, inv. no. Bo. 4327. Museum of the Great Mosque, Kaironan (*Arts of Islam* [supra n. 33] no. 177, "Egypt 11th C."). Example for nine conical lamps: Gemeentemuseum, The Hague (*JGS* 16[1974] 126, no. 9).
35. London, British Museum, O. M. Dalton, *Byzantine Art and Archaeology* (Oxford 1911) 622, fig. 395. A polycandelon for sixteen lamps, datable to the 7th C., is in the British Museum, inv. no. E.C. 529.
36. Cf. esp. V. H. Elbern, "Neuerworbene Bronzebildwerke in der Frühchristlich-Byzantinischen Sammlung," *BerlMus* n.s. 20 (1970) 8–11.
37. London, British Museum; cf. Elbern (supra n. 36) no. 26, and E. Cruikshank Dodd, *Byzantine Silver Stamps* (Washington 1961) no. 24.
38. Trans. W. R. Lethaby and H. Swainson, *The Church of Sancta Sophia Constantinople* (London and New York 1894) 50. Cited in Lamm, *Mittelalterl. Gläser* I 503–504. Cf. also Trowbridge, 190–191. For 9th C. polycandela and glass lamps see Lamm, *Samarra*, 32ff. Also wooden tripods as found in Egypt may have been used: Harden, *Karanis*, 155.

lamp. The lamps from Jerash seen by this writer in
the Ashmolean Museum in Oxford measure 10 to 12
cm. in height). The lamps recorded were predominantly fabric 1 and fabric 2.

The variant with long and plain or multiknobbed
lower part[39] does not seem to have been known at
Sardis. In contrast the lamps from Jerash—to cite
one important group—show greater variations in the
form of bowls and stems.[40]

The largest bowl section of a lamp, **275**, unfortunately comes from an unstratified, dump-like area at
HoB which contained everything "between 'late Hellenistic' and late Roman period" (*BASOR* 177, 14–
17). In analogy to Early Byzantine material listed in
the following catalogue entries, this portion must also
be datable to a time probably not earlier than the fifth
century. There is another section, **276**, from HoB
that was discovered in the top level of the Lydian
Trench.

Well-preserved lower portions of lamps, **274**, come
from an area east of Syn Fc where numerous fragments were found, among them particularly those of
goblets and bowl lamps. Hundreds of coins from this
area range from Constantine I to Heraclius; this material is, therefore, datable from the fourth to the early
seventh with concentrations in the fifth and sixth
centuries (*BASOR* 174, 47).[41]

Definitely Early Byzantine are thorn-like lamp
bases, **277**, from BS E 13 (BASOR 170, 51); **278**, from
Pa-S at a level associated with the Early Byzantine
period (*BASOR* 191, 28–29); **279**, from BS W 7
(*BASOR* 157, 32–33); **280**, from BS E 1 (BASOR 191,
17) found with salver rims **378** and **379** and sections
of bowl lamps, bottles, and windows. **281** was found
in BS E 17 which was particularly rich in Early Byzantine material, including the remains of goblets, bowl
lamps, pattern-molded ware, and windows (*BASOR*
174, 45).

From BE-H comes the lower part of a lamp, **282**,
found in the rubble beneath the floor. It is certainly
post-Roman (*BASOR* 177, 23); with it were various
vessel fragments of fabrics 1 and 3 of Early Byzantine
date. Another base, **283**, was found in 1963 at BE-A,
a room that was fully excavated in 1964 and 1966; a
rim and a concave vessel base in fabric 3 glass were
found with it (*BASOR* 187, 10, designated unit A'). At
RT were found two bases that can be associated with
coins mainly of the seventh century (*BASOR* 166, 44;
Sardis M1 [1971] 152, "Hoard M"). The first, **284**, was
with a goblet stem; the other, **285**, was with the section of a bowl lamp, **255**. Finally, **286**, from BE-B, is
similar to lamp bases of this type but did not necessarily form part of a lamp.

274 *Pls. 11, 23.* Lower portions of 3 lamps, funnel-shaped
with slender base. Fabrics 1, 2, and 3.
P.H. ca. 6, diam. at top ca. 4.2, of base 1.
E of Syn 1963 E116.7-E118.7/N5 *98.00-97.50.

275 *Pl. 11.* Waisted, with funnel-shaped lower (fractured)
part and upper body curving to bowl with vertical wall. Fabric 2.
P.H. 6.3, est. diam. rim 7, diam. of funnel 3.5; est. H. of
lamp ca. 8 (estimate based on combination of **274** and **275**).
MTE 1964 E40-E48/S120-S125 *101.40-101.00.

276 Fabric 1; badly made.
P.H. 5.2.
HoB 1962 W30/S95 to *100.40.

Bases

277 Several bases.
P.H. 3.3-4.3.
BS E 13, 1962 E75-E80/S0-S3.5 to *96.50 floor.

278 Fabric 2.
P.H. 5.7, diam. 0.9.
Pa-S 1967 E60-E65/N20.5-N22 *97.20-96.45.

279 G59.30:1390. Fabric 1.
P.H. 5.5.
BS W 7, W19-W23/S2 *97.50-97.10.

280 *Pl. 23.* Fabric 1.
P.H. 5.5.
BS E 1, 1967 E10-E12.50/S0-S5 *97.25-97.00.

281 Fabric 2.
P.H. 5.
BS E 17, 1963 E98.5/S1-S3 *94.00-93.50.

282 Fabric 2.
P.H. 3.5.
BE-H 1964 E11-E13/N40-N52 *96.80-96.50, below floor
level.

283 Fabric 1.
P.H. 3.2.
BE-A 1963 E12-E12.40/N0-N3 *97.50-97.00.

284 Base of large lamp. Fabric 2.
P.H. 2.5, diam. 1.5.
RT 1961 E4-E6/S5-S15 *96.50 down.

39. Crowfoot-Harden, pl. 29:21–25 (here *Pl. 23*).
40. Other lamps some of which have multiknobbed stems were
found in Hira, Iraq (Oxford, Ashmolean Museum), Saller, *Moses,*
317, pl. 140, 5th–6th C. and other sites.
41. *Sardis* M1 (1971) s.v. Findspot Index, Syn-Porch, p. 150. Ap-
pendix C, p. 158.

285 RT 1961 E16-E17/S9, on colonnade floor *96.50. Found with **255**.

286 Lamp base(?). Lower cylindrical tube-like portion of vessel. Fabric 1.
P.H. 2.6.
BE-B 1964 E24-E26/N1-N2 to *96.50.

Type 4

Examples of beaker-shaped lamps are less numerous than those of types 2 or 3. The largest portion preserved consists of the slightly convex wall section of a beaker-like vessel with a short base knob (preserved height 5.8 cm.).[42] Presumably the upper portion had a straight or slightly flaring rim. The total height may have been about 7 to 10 cm.; the base diameter of the Sardis finds varies from 1.3 to 2.7 cm.

The identification of these objects as lamps seems logical because they are unstable; they may have been suspended or placed into a metal lamp holder. Beaker lamps are closely related to handled bowl lamps and, in addition, resemble lamps still used in the Near East.[43] However, there appears to be no published example from a controlled excavation. A precursor may be a beaker lamp from Dura Europos.[44]

A curious fact often seems to hinder the investigation of many of the glass fragments in Sardis: the larger a vessel section of a given type happens to be, the less securely it is dated, and vice versa. Thus, sizable sherds such as **287**, **288**, and **289** come from unstratified or mixed levels and could, therefore, be Roman as well as Early Byzantine. Equally insecure is the date of the level at the northeast corner of unit C at HoB where **292** was found.

Fortunately the finds of lamps of type 4 from B come from better contexts. One base, **290**, discovered in 1968, from LNH is definitely Early Byzantine. Its fabric, almost clear, bubbly glass with faint green tint, is unusual. An Early Byzantine bottle neck of the same fabric was found in close proximity in 1966, **506**.

Lamp section **293**, from an area east of Syn Fc, is no doubt contemporary with the fragments of Early Byzantine beakers, goblets, and various lamp types found there. **294** was found inside the south wall of Syn with a ring base (cf. **401–444**) and the small handle of a bowl lamp (type 2).

The Byzantine Shops also contained beaker lamps: **295** from E 19, **296** from E 10, and **297** from E 17, that included much glass: remains of goblets, lamps, salvers, bottles, and windows. Finally a lamp base,

298, was found in the northwestern section of Pa, that differs slightly from the other examples as it is made of very thin glass and has a relatively small knob with a distinct constriction; however, it seems to be contemporary with the other lamps (*BASOR* 191, 33).

Lower Portions of Lamps

287 *Pl. 23*. G59.66:2189. Knob base (on which piece could perhaps stand), slightly convex wall. Olive tint (fabric 3?), relatively thick.
P.H. 4.8, diam. at top 5.3.
BS W 13, ca. W55-W58/S1-S5 *95.00, deep pit, unstratified.
AJA 66, 8, no. 8, pl. 8:8.

288 *Pl. 12*. Knob base, curving wall. Fabric 2.
P.H. 5.6, diam. at top 5.2, at base 1.3.
HoB 1965 *102.40-100.50, upper mixed fill.

289 *Pls. 12, 23*. Relatively pronounced knob base and convex, steep wall. Fabric 2.
P.H. 5.5; diam. at top 6, at base 2.
MTE 1964 E40-E48/S120-S125 *102.00-101.50.

290 *Pl. 23*. Colorless with faint green tint (fabric 2?).
P.H. 5.5; diam. at top 6, at base 1.9.
LNH 1968 E55.60-E63.85/N106.10-N106.16 *98.50-98.00.

291 Fabric 3.
P.H. 4.5.
BS E 1, E5/S0.8-S0.9 *99.60.

292 Fabric 3.
P.H. 3.3.
HoB 1962 W10/S85, S part, to *100.50.

293 Fabric 3.
P.H. 4, diam. base 2.
E of Syn 1964 E123-E125/N0.5 *96.90-96.80.

294 Fabric 3.
P.H. 3, diam. base 2.7.
Syn 1965 E80-E82, N side of N wall, *93.60-93.20.

---

42. When the first example of a beaker lamp from Sardis came to my attention (*AJA* 66, 8, no. 8), it was mistakenly thought to be Roman. Now, after some safely dated material has been recorded, there is no doubt that this type as found in Sardis is Early Byzantine. For late 4th C. lamps of practically identical form see Isings (supra n. 15) 262.2, fig. 1:3. Perhaps the "bottle" found in Corinth, dated to the 5th C. and having a lower body with knob, is a parallel to the Sardis finds: Davidson, *Corinth* XII no. 683, fig. 11. Cf. also Harden, *Karanis,* no. 466, pl. 16 (perhaps 4th–5th C.).
43. Crowfoot-Harden, 201–202, pl. 28:12.
44. *Terminus ante quem:* A.D. 256; Clairmont, no. 755.

295 Fabric 1.
P.H. 3, diam. base 1.7.
BS E 19, 1963 E112-E113/S0-S2 *97.50-97.00.

296 Fabric 3.
P.H. 3.5.
BS E 10, 1967 E62-E64/S0-S1 *96.21.

297 *Pl. 23*. Pale olive (fabric 3?).
P.H. 5, diam. base 1.6.
BS E 17, 1963 E98-E100/S1-S4 to *95.70.

298 *Pl. 23*. Differs slightly from the other lamps: the material is thinner, the knob smaller, with a more pronounced constriction. Fabric 1, eggshell.
P.H. 5.8, diam. base 1.0.
Pa-W 1967 E34.80-E43.80/N78-N85 *96.50-96.30.

Type 5

"Lamp type 5" may, or may not, represent a lamp. Even its date is uncertain as it is a chance find. The material (fabric 1) and its bell or funnel shape speak against a date before ca. A.D. 400. If it is a lamp, one could assume that it had a handle (or handles) attached to the upper portion, which have not been preserved, or an outsplayed rim to support it in the ring of a polycandelon. As the base is also missing, there is no way of telling whether the vessel has a foot or a pointed base. Should the latter prove to be true the object might have resembled a type of lamp found in S. Menas (between Alexandria and the Wâd î Natrûn), a sanctuary" flourishing from the fifth to the eighth century."[45]

299 *Pls. 12, 23*. One half of vessel, perhaps a lamp or a footed bowl; bell-shaped body with waisted upper part, tapering to bottom (which, if fragment was a bowl, may have had a stem and foot). Fabric 1, eggshell.
P.H. 7, est. diam. rim 14.
Chance find.

GOBLETS

Drinking vessels in the form of goblets appear to have been among the most frequent vessel types in Early Byzantine Sardis. Fragments of at least 500 stems and feet, securely identified as goblets, were found.[46] No doubt this number should be doubled from the vast number of unidentifiable fragments. Almost all of the goblets come from Early Byzantine levels.

Unfortunately not one single goblet was preserved in a condition that shows a continuous profile. By analogy to material found elsewhere, the bowl form of a typical goblet from Sardis may have been slightly conical, U- or bell-shaped, while the rim seems to have been plain.

The stems occur in four major variants:

1. plain, short, slender, and slightly concave (waisted)
2. with knob that can vary from a pronounced central swelling to a ball
3. multiknobbed
4. pin-like

The feet also occur in four forms:

a. sloping with plain edge (most common)
b. sloping with folded edge
c. domed with plain edge (rare; cf. footed bowls, **382–383**)
d. flat and solid

Stems and feet appear in the following combinations; listed in approximate order of frequency:

1d. plain stem with flat foot
2a,2b. knob stem with sloping plain or folded foot
1a. plain stem with sloping plain foot
1c. plain stem with domed foot
1b. plain stem with sloping folded foot
4a. pin-like stem with sloping plain foot
3. multiknobbed stem, probably with flat foot having a tooled, profiled top

At Sardis, plain stems in a number of fabrics, are most frequently combined with various types of feet. Although they vary in technical execution, they resemble each other closely and stems of groups 1a, 1b, and 1c are indistinguishable without their feet.[47]

45. Crowfoot-Harden, 203, pl. 29:27 (here *Pl. 23*). A bell-shaped lamp with engraved grooves, in form similar to the example from Sardis, was published by Dusenbery (supra n. 11) 23, no. 37, fig. 36 (see also Sotheby Parke Bernet, New York, sale [Dec. 14, 1978] no. 107; *Apollo* 108 [1978] 211). Cf. a lamp roughly in the shape of a broad and low bell, having a pointed base and small handles attached to a rim with applied threads: Museum für Kunst und Kulturgeschichte, Dortmund, Schloss Cappenberg. Cf. also a more shallow, bell-shaped lamp in Dumbarton Oaks: Ross (supra n. 12) 59–60.

46. Some of the stems may belong to salvers, footed bowls (cf. **374–383**) or even lamps. For bowl lamps with feet cf. Barag, "Glass Vessels," pls. 28, 40.

47. When a few stems and feet of a particular combination—e.g. 1a—were found at a specific location, other stems without feet that came from the same spot were added to this group, i.e. to 1a, in the

The best fabric and the greatest technical skill are evident in groups with slender and elegantly curving stems resting on sloping or domed feet (1a, b, c). They are made of pale aquamarine (fabric 1) and pale olive (fabric 3) glass, in a ratio of about 5:1. The goblets with knobs (2b), though usually well-made, do not seem to reach the high level of craftsmanship evident in the former groups. Badly preserved examples of undeterminable color are most probably also manufactured in fabrics 1 and 3. Only very few are made of clear (fabric 5) glass. All of these variants are carefully finished, irregularities are rare, and the fabric is thin and free of bubbles and stones.

In contrast goblets with plain but less elegant, clumsier stems supported by flat feet (1d) are generally irregular, showing relatively poor craftsmanship. They are more often made of pale olive (fabric 3) than of aquamarine (fabric 1) glass and tend to decompose easily; many are covered with heavy weathering and frequently display dirty fire scum. It would be most interesting to determine whether the fine ware preceded or succeeded the coarser, whether the two were made concurrently in the same factory, or were made concurrently in different workshops, or whether all these are possible.

In BS E 13 (*BASOR* 174, 45), for example, quantities of goblet fragments—and other vessels—were found, including types 1a, 2b, and 4. This suggests that these variants were made during the same period, which might lead one to conclude that most of the Early Byzantine glass vessel types found at Sardis were made concurrently, be it in one single factory or in different workshops.

There are a number of slight variations resulting from individual finishing operations. The knob stems, in particular, tend to occur in different forms. Some are ball-like and sometimes striated while others have less pronounced swelling. Feet do not appear to have been made systematically with, or without, folded edge. It seems more likely that one workshop or individual glassmaker preferred folded feet while in another shop, or period, feet with plain edges were popular. However, one cannot exclude the possibility that both variants were made in a single workshop. Quite a few "unique" pieces will prove to have been the result of a glassmaker's whim rather than the conscious wish to create a new type. On the other hand, such pieces might also be the sole witnesses of important vessel groups not yet identified.

In the catalog listing the goblets are grouped according to stem formation. Although not one single piece is preserved in its entirety, one can estimate the average height of a Sardis goblet. One, **319**, is pre-

served to a height of 7 cm.; the multiknobbed stem, **367**, is 4.4 cm. high. Both may originally have measured ca. 8 to 10 cm. They seem to represent a median height while some large ones may have reached 13 to 15 cm. The diameters of the feet range from 4 to 5.4 cm., with a median diameter of 4.4 to 4.8 cm. The estimated diameter of an average goblet rim may have varied from 6 to 10 cm.

Goblets of blown glass with conical, U-shaped, or bell-shaped bowl and short stem either plain or in the form of a knob begin to appear in the first century A.D.[48] Their popularity increased during mid- and late Imperial periods.[49] Those without mold-blown or applied decoration, generally having a conical bowl, short concave or knobbed stem, and sloping foot with plain or folded edge, became the favored goblet form in the East and in Western regions, for example, in Italy.

Most closely related to the Sardis finds are goblets from Jerash and other places in Syria. The glass from Jerash appears to range from the fourth to fifth century; among the objects are goblets practically identical to our types 1 and 4.[50] Similar glass from the Ital-

statistics. As we used this method consistently when computing fragments for statistical purposes, the margin of error should be small.

48. Isings, no. 40. Cf. also Berger, 38–40; Saldern, *Boston*, no. 46.

49. Cf. Clairmont, nos. 460ff.; Harden, *Karanis*, 167ff., nos. 479ff. (4th–5th C.). Various goblets of perhaps Syrian (Palestinian) provenance and of late Roman date are included in Dusenbery (supra n. 11) 22, nos. 33ff. For numerous bases of typical late Roman goblets (4th C. and perhaps later) see G. D. Weinberg, "Glass Manufacture in Ancient Thessaly," *AJA* 66 (1962) 133, pl. 28, fig. 16; Bagatti-Milik, 144ff., fig. 34: 23–25. S. M. Goldstein, "Glass Fragments from Tell Hesbân, A Preliminary Report," *Andrews University Seminary Studies* 14 (1976) 130, pl. 11:D. Late Roman goblets from Sucidava (Romania) are decorated with applied and pincered thread bands and small handles (the latter similar to those of the bowl lamps in Sardis): Tudor (supra n. 8) esp. 88, fig. 31. One goblet from Ophal with deep bowl and knobbed stem, the bowl having an applied blue thread, is in the Palestine Archaeological Museum, Jerusalem (inv. no. 1751). Another goblet has a deep bowl with pattern-molded ribs and high stem with collar, in the Victoria and Albert Museum, London, apparently from "Egypt, 5th–8th C."

50. Some goblets from Jerash have an estimated height of ca. 8–9 cm.; the diameters of the feet measure about 4.5 cm. I am most grateful to the authorities of the Ashmolean Museum, Oxford, for having permitted me to study the finds from this site. Cf. Baur-Kraeling, 524ff., fig. 19–21. A large knobbed stem of uncertain date (5th–7th C.?): V. H. Elbern," Le pied de verre d'un vase liturgique(?) trouvé à Qâlât Semân," *Annales du 3e Congrès des "Journées Internationales du Verre," Damas, 1964* (Liège [1965]) 99–103. Comparable goblets are from Apamea, Canivet, 64–66, figs. 10–13 6th C.); Shavei Zion, Barag, *Shavei Zion*, 67, nos. 15–17,

ian and Southeast European excavations is proof of the international style of this goblet type.[51] The goblets discovered in Islamic contexts in the Near East and in Corinth do not differ basically from the finds from Karanis or Sardis. Those of the Islamic sites date from the ninth to tenth century,[52] while the material from Corinth is datable to the early eleventh to mid-twelfth century.[53]

Two forms from Corinth are actually direct successors of types found at Sardis: the flat-footed variant with plain stem[54] follows the Sardis type 1d, and the knobbed goblet,[55] Sardis type 2b. Thus, the manufacture of almost identical vessels in various parts of the ancient world in late and post-Roman times again shows the conservatism inherent in the production of utilitarian glass.

Types 1a, 1b, 1c

This group comprises goblets with slender concave stem and a foot which usually has a plain, occasionally a folded, edge; the foot is rarely domed. Most were found with the remains of other goblets (particularly type 2), glass vessels, and window panes. They almost always come from safely dated Early Byzantine levels (predominantly the Byzantine Shops), covering a period from about the early fifth to early seventh century, with an apparent preponderance of finds datable to the sixth and early seventh century.

BS E 13, in which **300** was found, contained the largest deposit of fragments of glass vessels; a rough count came to about 2,000. With **300** were at least twenty similar stems, mainly of aquamarine (fabric 1) and clear (fabric 5) glass, over a dozen knobbed stems (type 2) many of which are also of fabric 1, and pin-like stems (type 4). In addition, there were masses of bottle sections, vessel rims, and window glass.

The level under Syn Apse where **301** was found predates the Early Byzantine period (*BASOR* 191, 26). However, as it is identical to the other examples of this type it may be an intrusion. Two stems, **302** and **303** were found together in BS E 17 with knobbed stems (type 2b), the remains of goblets with flat feet (type 1d), and five ring bases (cf. **401–444**; *BASOR* 174, 45–46).

A number of goblet fragments (**304–309**) were discovered when cleaning Syn Fc and the area to the east (*BASOR* 174, 43–44; 177, 19). Knobbed, flat-footed, and pin-like stems were also excavated there. The levels are associated with Early Byzantine material although no *terminus post quem* could safely be established. The whole group probably belongs to the fifth to sixth century.

fig. 16 (5th–6th C; here *Pl. 23*); Mt. Nebo, Saller, *Moses,* 318–319, pl. 140 (ca. 5th–6th C.); Khirbet el-Kerak, Delougaz-Haines, pl. 60:14–23 (mainly early 5th-early 7th C.); Samaria, Crowfoot, *Samaria,* 414ff., fig. 96 (the remains of over 40 goblets were found in a ca. 4th–6th C. context); Beth Shan, FitzGerald, 42, pl. 39; Auja Hafir, Harden, *Nessana,* 86, no. 62, pl. 20, with thin stems similar to Sardis, type 4a; and other places, cf. esp. the goblets from Mount Olive, Jerusalem (ca. late 4th C.): Bagatti-Milik, 145–146, fig. 34:23–24; Barag, "Glass Vessels," pls. 10, 22, 23, 26, 28, 33. Puttrich-Reignard, 20ff. (ca. 5th–6th C.); goblet stems with flat feet and convex stems (similar to Sardis type 1d) from Ctesiphon are preserved in The Metropolitan Museum of Art, New York. Similar material was found in Kish, datable also to the 5th–6th C. (S. Langdon and D. B. Harden, "Glass from Kish," *Iraq* 1 [1934] fig. 5:16), Sarachne, 6th–7th C. (mentioned in Harden, *ArchJ,* 83, n. 14), and Babylon (cf. Puttrich-Reignard, 21). Renate Pirling kindly brought to my attention goblets of the 5th–6th C. found at Apamea in Syria; they represent about one-third to one-half of the total number of glass finds and come in various shades of green and bluish-green.

51. They include the materiál from Rome, Isings (supra n. 15) 262.2–3. Castelseprio, 6th C.: Leciejewicz et al., 162ff., fig. 5. S. Kurnatowski, E. Tabaczynska, and S. Tabaczynski, "Gli scavi a Castelseprio nel 1963," *Rassegna Gallaratese di Storia e d'Arte* 27 (1968) 75, fig. 6. For this and the following sites cf. Philippe, 94ff. Torcello, 7th–8th C.: remains of goblets with flat feet and concave or almost cylindrical stems, close in style to Sardis goblets type 1d. Cf. L. Leciejewicz, E. Tabaczynska, and S. Tabaczynski, "Ricerche archeologiche nell'area della cattedrale di Torcello nel 1961," *Bollettino dell'Istituto di Storia della Società e dello Stato Veneziano* 3 (1961) 28–47. Idem, "Ricerche archeologiche a Torcello nel 1962, relazione provisoria," ibid. 5–6 (1963–64) 3–14. A. Gasparetto," A proposito dell'officina vetraria Torcellana," *JGS* 9 (1967) 50–75. Invillino, ca. 5th–7th C.: remains of goblets with stems and feet identical or very similar to Sardis goblets of types 1 and 2, Fingerlein et al., 73ff., fig. 13; Philippe, fig. 50. Nocera Umbra: another type of goblet with opaque white threads marvered into the surface and a rim turned outward is represented by a goblet from this site, datable to the late 6th or 7th C.; cf. A. Pasqui and R. Paribeni, "Necropoli barbarica di Nocera Umbra," *MonAnt* 24 (1918) fig. 20; Harden, *ArchJ,* 85, fig. 4. Cf. also Clairmont et al. (supra n. 15) 164–165, pl. 31 (late Roman). More goblets of the late 5th–6th C. were found at Iatrus near Krivina (Bulgaria), having bell and U-shaped bowls (information kindly supplied by Gudrun Gomolka).

52. E.g. A. Lane, "Medieval Finds at Al Mina in North Syria," *Archaeologia* 87 (1938) 63ff., fig. 10:Q, R. Cf. also M. Negro Ponzi, "Islamic Glassware from Seleucia," *Mesopotamia* 5–6 (1970–71) 91–93, nos. 111ff., fig. 54 (with additional ref.). Goblets of a different form, having a conical bowl with a wide collar at its base, apparently represent a separate development in the Near East; they seem to be of 8th and 9th C. date; cf. Saldern, *Jb. Hamburg* 57–59, fig. 15; idem (supra n. 3) 59–60, figs. 4–5. Cf. also P. J. Riis, "Les verreries," in V. Poulsen et al., *Hama. Fouilles et Recherches 1931–1938. IV.2. Les Verreries et Poteries Médiévales* (Copenhagen 1957) 44, figs. 92–93. A goblet (or lamp) with conical bowl and a collar, having three little handles at the rim and supported by a knobbed stem, in West Berlin, Islamisches Museum. A goblet with bell-shaped bowl on a high, knobbed stem recently sold in London was dated to the 14th–15th C. but seems to be earlier, i.e. late first millennium A.D.: *The Constable-Maxwell Collection of Ancient Glass,* Sotheby's (June 4–5, 1979) no. 355.

53. Davidson, *Corinth* XII nos. 712ff., fig. 12; G. D. Weinberg, "A Medieval Mystery: Byzantine Glass Production," *JGS* 17 (1975)

The Byzantine Shops excavated in 1958 and 1959 and the area west of B directly adjacent to them—including the latrine (*BASOR* 154, 16–18; 157, 32–33)—revealed quantities of glass vessels and window panes, including **310–311**.[56] As noted before, vessels of blue glass are extremely rare in Sardis; two goblet sections of this fabric 6 were found in the Byzantine Shops in 1959, **315** and **316**.

In 1966 a goblet section, **317**, came to light in BE-B (*BASOR* 187, 15ff., designated room B'); although this room has a complicated history (*BASOR* 187, 12) the level in which the goblet fragment was found is probably datable to the fifth or sixth century. Other material from this level includes ring bases (cf. **401–444**) and window glass. Another example, **318**, comes from BS W 14 (*BASOR* 186, 29); there is no doubt that it should be associated with the fifth or sixth century. Quantities of window panes were found in this shop.

Finally three examples of goblets of type 1 may suffice to represent the finds from HoB. Two were discovered in 1959 in area 11 A (**319**) and 5 (**320**) in Early Byzantine levels, datable mainly to the sixth and early seventh century (*BASOR* 157, 24); **320** is particularly interesting as it was found in a niche in the north wall of area 5 that also contained the bell beaker **472**, and the salver **374**, two of the very few glass finds from Sardis that are fairly complete.

Type 1d

The goblet with flat foot is probably the most common type at Sardis. In general it is less carefully made than the other variants. Although the stem lacks the elegance of type 1a, both series are closely related. The foot is flat—not sloping—and usually not perfectly circular. When only the stem is preserved one can easily determine whether it belongs to type 1a or 1d: the latter is a clumsier, and perhaps less expensive, variety of the more refined 1a. Both types were certainly made during the same period, as examples of both are often found together.

Most are made of light green glass (fabric 2), in contrast to types 1a–c which are predominantly of fabric 1 glass. The finds tend to be less well-preserved; many of them are heavily weathered and show dirty fire scum. The size of a goblet must have been similar to that of types 1a–c.

The dating of the goblets concurs with the dating of the other Early Byzantine material; most finds come from the Syn and B areas.

In BS E 13 (*BASOR* 170, 50; 177, 19) at least one goblet foot, **323**, was found with goblets of types 1a

and 2, as well as with lamps, pattern-molded ware, ring bases (cf. **401–444**) and other typically Early Byzantine material. Goblet sections were excavated at Syn Fc, **324**, BS E 17, **325** (*BASOR* 174, 45–46; although found as deep as *94.00 it certainly belongs to the Early Byzantine period), and at the area east of Syn where many goblets of other types came to light, **326–327**. In addition, fragments of goblets were found in unit BE-B, **328** (in a level later than A.D. 400 but before the mid-seventh century; *BASOR* 187, 16, there unit B') and in an area west of B (W of B-W), **329**, which, at this level, primarily contained material which was earlier than A.D. 616 (*BASOR* 191, 38–39, Trench E).

None of the goblet sections from HoB, **331–333**, can be dated safely as they come from disturbed levels. The fragments from PN, **334–336**, were also not well stratified. **335** was from a stratum associated with the Roman period (*BASOR* 174, 22ff.). It was found near an engraved bowl fragment, **45**, and the rim fragment of a cylindrical beaker with engraved horizontal lines, **47**, both datable to the third century. Indistinguishable from the Early Byzantine glass, this goblet fragment may be an intrusion or a predecessor of the Early Byzantine goblets.

Fragments

Predominantly fabric 1, eggshell, slightly weathered.

300 P.H. 2, diam. foot 5.
BS E 13, 1962 E76-E79/S2.5-S3.5 *96.40-96.10.

301 *Pl. 24.* Diam. foot 4.2, H. of stem to lower bowl 2.
Syn 1967 E36-E39.60/N11-N12, apse trench *92.50-92.00.

302 BS E 17, 1963 E97-E101/S0-S4 *95.20.

303 BS E 17, 1963 E99-E102/S2-S4 *97.50-97.00.

304 H. stem 2, est. diam. rim 7.
Syn Fc 1963 E115-E118/N3-N5.5 *98.00-96.75.

305 Group of 9 goblet stems and feet.
Syn 1963 E113-E115/N6-N8 *97.50-96.75.

138ff. Cf. also D. B. Harden," Some Glass Fragments, Mainly of the 12th–13th Century A.D., from Northern Apulia," *JGS* 8 (1966) 74ff.

54. Davidson, *Corinth* XII nos. 711–718.

55. Ibid. nos. 720–722.

56. Hanfmann, *JGS,* 53; *AJA* 66, 9–10.

306   E of Syn 1964 E125-E128/S1-N1 *98.00.

307   *Pl. 24.* Greenish, very thick.
H. stem 0.5, diam. foot 3.8.
E of Syn 1963 E117-E121/N2-N4 *97.75-96.75.

308   *Pl. 24.* Stem (of goblet?). Fabric 4, comparatively
heavy.
P.H. 3.3, diam. stem 1.2.
E of Syn 1964 E123-E126/N6-N7 *97.00-96.00.

309   Stem (of goblet?). Fabric 2, heavy white scum, foot
fire-softened.
P.H. 3, diam. stem 1.
E of Syn 1963 E118-E119/N0.1-N2.00 *97.50-96.75.

310   G58.34:528.
BS W 2, W 8, 2.50 below top of S wall corner, level III.
Hanfmann, *JGS,* 53.

311   G58.44:586, G58.45:587, G58.46:611.
BS W 1, W1-E2/S2-S4 *96.40.
Hanfmann, *JGS,* 53, fig. 2.

312   G59.29B:1382.
P.H. 2.8.
BS W 4, not well stratified.
AJA 66, 10, no. 10e, pl. 9:10e.

313   G59.67:2193.
P.H. 2.7.
B W alignment, latrine floor.
AJA 66, 10, no. 10f, pl. 9:10f.

314   *Pl. 12.* G73.7:8264.
P.H. 3.2, P.W. 4.7.
PN/E on top of grave 73.13, W214/S359 *90.30.

315   Small fragment of vessel, most probably a goblet. Fabric 6, blue.
BS W 13, 1959 W55-W57/S1.70-S4.40 *98.50-97.00.

316   Small section of folded foot of goblet. Fabric 6, blue;
inner "lining" of tube-like cross section has disintegrated to
white.
BS W 13, 1959 W52-W54/S2.50-S4.40 *96.50-95.50.

317   BE-B 1966 E24-E27/N4-N6 *96.80-96.40.

318   Diam. foot 4.6.
BS W 14, 1966 W60-W68/S4-S7 *98.00 down.

319   *Pl. 12.* G59.37:1456.
Diam. foot 5, of bowl at fracture 7; P.H. 7.
HoB area 11A, in drain in E wall.
AJA 66, 9, no. 10a, pls. 6:13, 9:10a; Hanfmann, "Décou-
vertes," 124–125, fig. 78.

320   G59.42a:1769. Portions of lower bowl, short stem and
domed (?) foot.
Est. H. ca. 12-15, est. diam. rim ca. 10.
HoB area 5, niche in N wall. Found with **374** and **472**.

321   Sloping foot, slender stem, spreading bowl; well-
made. Fabric 1.
P.H. 3.5, diam. foot 5.
HoB 1959 *97.40-96.70.

322   Pale blue, bubbly.
MMS 1978, E147-E149/S55.70-S57.50 *102.50-101.95

Feet and Stems

Predominantly fabric 2, heavily weathered, some
pieces with dirty fire scum.

323   *Pls. 12, 24.* P.H. 3, diam. foot 4-4.9.
BS E 13, 1964 E75-E80/S0-S3.5 to *96.50 floor.

324   Syn Fc 1962 E97-E99/N1-N3 *97.00-96.65.

325   BS E 17, 1963 E98.5/S1-S3 *94.00-93.50.

326   E of Syn 1963 E120-E122/N1.5-N2 *97.50-96.75.

327   E of Syn 1964 E128-E131/N4-N8 *97.90-97.80.

328   BE-B 1966 E21-E25.5/N6-N9.9 *97.00-96.50.

329   W of B-W 1967, trench E, W76.50-W81 S1.50-N4.25
*97.60-97.30.

330   MRd 1961, S of colonnade E9-E15/S12-S15 *96.70(?).

331   HoB 1965 W5-W10/S115-S118 *99.15-98.90.

332   HoB 1963 E10/S120, to *101.75.

333   HoB 1964 W10-W18/S171 *106.30-106.10.

334   PN 1964 W250-W253/S340-S350 *88.50.

335   PN 1964 W258-W259/S341-S343 *88.50-88.10.

336   *Pl. 24.* Irregular, corrugated flat foot, on which a few
pieces of glass applied or dripped and then flattened. Fabric
1; badly made.
P.H. 3, diam. foot 4.5.
PN 1963 W225/S339-S342 *88.40-88.00.

Types 2a and 2b

Goblet stems with knobs or a bulge were perhaps as
frequent as those of types 1a or 1b. The feet have either plain or folded edges. Types 1 and 2 are fre-

quently found together and, therefore, were manufactured concurrently. Height and diameters are identical or very similar to those of type 1.

A typical, well-preserved section, **337**, comes from the area east of Syn Fc where much Early Byzantine material—including goblets of type 1—was found; however, the stratification is not very clear (*BASOR* 174, 43–44). Another stem from about the same area, **338**, is deformed by heat; it might be associated with the rebuilding of the fortifications near the Byzantine tower in the sixth century (*BASOR* 177, 19).

Two stems, **340**, from BS E 15 (*BASOR* 174, 45) and **341**, from near the apse of Syn (*BASOR* 174, 30ff.) are associated with Early Byzantine levels. In BE–S, **346** was found with ring bases (cf. **401–444**), and from BS E 13 are **347** and **348** (*BASOR* 170, 51).

The knobbed goblet stem sections from HoB (**349–351**) were found in Byzantine levels of the sixth and early seventh century, overlaying the Roman and Hellenistic occupation (*BASOR* 157, 24; 177, 13–14).

At PN, **352**, was found among the many glass fragments in the late Roman-Early Byzantine bath, provisionally dated to the fifth century; it was discovered close to sections of bottles (**487** and **518**) and a bowl lamp section (**237**).

A number of fractured knobs without feet and lower portions of bowls probably belonged to goblets of these types although some may have been part of footed bowls or salvers. They come from: Syn Fc (**355**), where goblet fragments were common; from the area east of Fc, (**356**; fragments of blue glass and mosaic tesserae were found close-by); from BS E 16, with at least six other goblet feet, a flat cup base (**446**), and window glass (*BASOR* 174, 45); from various locations at B (**357–360**), and from the dump north of LNH (**361**; *BASOR* 191, 33). Two examples are from the Islamic-Byzantine level at PN; one, **364**, was about 1 m. below an Islamic occupation that contained coins of Suleyman II (1687–1688; *BASOR* 177, 3).

Feet and Stems

Fabrics 1 and 2, most heavily weathered, some pieces with dirty fire scum.

337 Well-preserved.
P.H. 2.6, diam. foot 4.1.
E of Syn 1963 E115-E118/N3.5-N5 *98.00-96.75.

338 G64.8:6520. Misshapen by burning.
P.H. 4.5, diam. foot 4.
E of Syn E128-E130/N5-N8 *97.00-96.00.

339 Syn 1962 E116-E118/N3-N5, in angled walls *97.58-97.00.

340 BS E 15, 1963 E88-E93.5/S1, excavation drain *96.10-95.30.

341 Syn 1963 E38-E43/N4-N8 *97.80-96.75.

342 BS E 11, 1967 E69-E71/S0-S1 *96.20.

343 *Pl. 24.* Knob of stem of large vessel.
P.H. 2.5, diam. 3.4, Th. up to 0.5.
BS E 8, 1968 E53-E59/S5-S5.50 ca. *96.80.

344 Syn 1967 E38-E43/N4-N8 *97.80-96.75.

345 BS E 1, 1967 E6-E8/S3-S6 *97.05.

346 BE-S 1965 E20-E22.5/N29.75-N36 *96.35.

347 BS E 13, 1962 E75-E80/S0-S3.5 to *96.50.

348 BS E 13, 1962 E75-E76/S2-S9 *96.70-96.20.

349 HoB 1964 W20-W25/S115-S119 *99.15-98.95.

350 G59.57:2001. Fabric 3?
P.H. 2.9.
HoB room 7, floor 2 *96.77.
AJA 66, 10, no. 10c, pl. 9:10c.

351 *Pl. 24.* G59.40:1480.
P.H. 3.1.
HoB area 15.
AJA 66, 10, no. 10d, pl. 9:10d.

352 PN 1961 W265/S360 ca. *89.10.

353 PN 1965 W285-W290/S320-S325 *87.20-86.70.

354 PN 1965 W303-W307/S323-S327 *85.90-85.70.

355 Syn Fc 1963 E104-E108/N1-N2.5 *97.30-97.00.

356 Syn 1963 E122-E125/N8-N10 *98.00-96.50.

357 Vault under MC 1964 E33.50-E35.00/N57-N59.50 *91.97.

358 Tunnel under MC 1964 E33.50-E35.10/N57-N58.50 *94.18-93.92.

359 W of B 1966 W62.5-W64.8/N0.38-S2.20 *96.60-96.00.

360 BWH 1961 W43-W55/N61.5-N68.5 *101.00.

361 Dump NE of LNH 1967, ca. *98.50.

362   RT 1961 E13-E15/S25-S27 *96.75-96.30.

363   BE-B 1966 E17-E26/N0-N9 *100.20-98.50.

364   PN 1964 W260-W263/S344-S348 *87.50-87.30.

365   PN 1964 W251-W255/S347-S350, inside wall E, *89.35-89.00.

366   HoB 1962 E20/S90 to *100.30.

Type 3

Stems with two or more knobs in consecutive series seem to have been introduced in the first century A.D.[57] The material datable to Early Byzantine and Islamic periods, however, probably represents a development from vessel stems with a single knob. Among the multiknobbed stems which are close to the Sardis finds, are vessels used as lamps found at Jerash and other sites[58] and goblets with enameled decoration from Raqqa which seem to be of the ninth century.[59]

Only one complete stem was found in Sardis, **367**, which was probably part of a goblet but could conceivably have belonged to a lamp; its foot has a profiled top. The preserved height of the stem is 4.4 cm. which may represent about half the total height of the goblet. The bowl may have been wide and conical or have had a straight, vertical wall. The rarity of this type in Sardis might be an indication that **367** was an import, although this seems most unlikely. The stem was discovered in an area where much Early Byzantine glass came to light, namely in BS W 13 close to the public latrine (*BASOR* 157, 34). At least seven sherds of similar stems were found in BS E 12 in 1964. A double knob from a stem, **369**, discovered in 1967 in BS E 4, may have belonged to a similar goblet.[60] In this shop were found sections of pattern-molded lamps (**239**) as well as fragments of vessel walls of fabric 1 and window glass.

367   *Pl. 12*. G59.68:2194. Stem formed of 5 knobs. Portions of foot and lower bowl preserved, the foot with concentric coil-like ribs. Fabric 2; heavy weathering.
P.H. 4.4.
BS W 13, floor level.
AJA 66, 10, no. 10g, pls. 6:14, 9:10g.

368   Portions of 7 multiknobbed stems.
BS E 12, 1964 E73-E74/S1-S2.5 *97.50-96.50.

369   One and one-half knobs of a stem, presumably from a goblet. Fabric 1.
P.H. 2, diam. 1.8.
BS E 4, 1967 E29-E33/S0-S6 *97.90-97.10.

Type 4

A few fragmentary goblets were found that are related to type 1a but have an extremely thin, pin-like stem resting on a sloping foot. This stem form seems to be a distinct and consciously developed variant. Bowl form and height must have approximated that of the other goblets. Two are greenish, probably of fabric 2; the others are covered with heavy iridescence, making it impossible to identify the fabric.

The closest parallels are a few fragments of goblets found by Thompson in a Sassanian context in Nineveh in 1928-29.[61]

57. Isings, no. 36 a.
58. Crowfoot-Harden, nos. 21-23; Baur-Kraeling, 519ff., fig. 17 (4th-5th C.). Goblets with two knobs, datable to the 5th-6th C., were found in Ctesiphon; Puttrich-Reignard, 21. The problem of whether vessels of this type were used as goblets or lamps—or both—cannot be solved at present. Among the vessels with multiknobbed stems are: a conical goblet with a band of arches incised with a sharp tool in the hot glass, in the West Berlin, Antikensammlung; *JGS* 4 (1962) 66, fig. 12 (perhaps datable to the 5th-7th C. For the technique of producing grooves in the hot glass cf. *Smith Coll.*, no. 425). A multiknobbed stem ca. 8 cm. high, fractured at the lower bowl which may have been conical, is in the Persepolis Museum (without archaeological record). For a goblet or lamp closely related to these vessels, see Wulf (supra n. 31) no. 1004. A vessel with broad conical bowl having a flange and supported by a stem with three knobs, perhaps of the late 4th C., cf. N. Avigad, "Excavations at Beth She'arim, 1954. Preliminary Report," *IEJ* 5 (1955), pl. 33:1.
59. S. Abdul-Hak, "Contribution à l'étude de la verrerie musulmane du VIIIe siècle," *Annales du 1er Congrès des "Journées Internationales du Verre," Liège, 1958* (Liège n.d.) 89-90, fig. 15. Equally of 8th C. date: M. Negro Ponzi (supra n. 52) 92, nos. 114ff. For a goblet of about this date with diamond-point engraved decoration cf. C. W. Clairmont, "Some Islamic Glass in the Metropolitan Museum," *Islamic Art in The Metropolitan Museum of Art*, ed. R. Ettinghausen (New York 1972) 145-146, no. 8. Vessels of Islamic date consisting of deep bowls with mold-blown decoration, resting on multiknobbed stems are preserved in West Berlin, Islamisches Museum, and in the Benaki Museum, Athens.
60. Or possibly a handled lamp of the type Crowfoot-Harden, pl. 29:21-23 (here *Pl. 23*), from Jerash. One example in Oxford, Ashmolean Museum, measures 7.3 cm. up to the fracture of the bottom of the bowl.
61. Oxford, Ashmolean Museum ("B 27, K XXIV 3, below first cement"); cf. D. B. Harden, "Notes on the Glass Found During the Excavations," in R. Campbell Thompson, M. E. L. Mallowan, "The British Museum Excavations at Nineveh, 1931-1932," *AnnLiv* 20 (1933) 184-186. Cf. also Harden, *Nessana*, 86, no. 62, pl. 20: relatively thin stem of a goblet (or cup?), datable perhaps to the 4th-7th C.

The find pattern of the Sardis goblets is similar to that of the other types. **370** comes from southeast of Syn (*BASOR* 177, 19). A group of about twenty sherds of goblets of this type were found in BS E 13 in 1962 with a goblet of type 1a, **371** (*BASOR* 170, 51). Another fragment, **372**, comes from unit BE-A and must date between the early fifth century and the destruction in A.D. 616; in its immediate vicinity was a small handle probably of a lamp of type 2 (*BASOR* 187, 12–14). Finally a goblet section was found in 1965 at the entrance of unit BE-A, between Shops W 1 and E 1, **373**.

370 Fabric 2; highly iridescent.
Diam. foot 3.5.
Syn 1964 E125-E127/S10-S12 *97.00-96.50 Southeast of "Byzantine Tower" (*BASOR* 177, 19).

371 BS E 13, 1962 E76-E79/S2.5-S3.5 *96.40-96.10.

372 BE-A 1966 E5.40-E8.10/N9-N10.80 *97.30-96.80.

373 BE-A entrance 1965 E1-E3/S0-S3 ca. *96.75.

SALVERS

A fairly complete salver, **374**, was found with a beaker, **472**, and a goblet foot, **320**, in a niche of unit 5 at HoB; the archaeological context is datable to the sixth or early seventh century (*BASOR* 157, 24–26). The salver is made of fabric 1 glass and consists of a shallow, wide bowl with a diameter of 12.5 cm., resting on a flaring foot with its bottom edge turned downward. As far as we know it is the only object of this type and date yet published that comes from a controlled excavation. Similar vessels may have been found in Ctesiphon.[62] The salver shape derives from shallow bowls and plates common in mid-Roman Imperial times[63] in Egypt and Syria.[64]

The foot construction is similar to that of goblets found at Sardis. More than one fragment of a foot identified in the catalogue (cf. **300–373**) or in the statistics as having belonged to a goblet may actually have been part of a salver. The identification of salver rim fragments is even more difficult. By analogy with the salver from HoB (**374**), rims of this group tended to be plain and were turned downward to an almost horizontal position, markedly different from the straight goblet rims. However, as some goblets appear to have had slightly flaring rims, salvers and goblets could easily be confused.

Two feet and several rim sections appear, because of their size, to have formed parts of salvers. One

foot, **375**, was found in BS E 16 (*BASOR* 174, 45); another, **376**, comes from Pa-W (*BASOR* 191, 41). The former, of sixth century date, was found with the section of a conical bottle neck similar to **482**. The latter, associated with the mosaic floor, is probably also datable to the same period; with it came a bottle neck with spiral thread (**609**), a plain bottle neck, and various stems of either goblets or bowls (**384** and **385**), objects typical of the sixth and early seventh centuries.

According to the estimated diameter and angle of inclination, the rims listed in the following may also have been part of salvers. They are, with one exception, of fabric 1, and their estimated diameter ranges from 13 to 15 cm. One, **377**, was found in BS E 13, with fragments of lamps, goblets, bowls, and bottles, all datable to the fifth to sixth century (*BASOR* 170, 51). Two others, **378** and **379**, were from BS E 1 which contained sections of lamps, bottles, and windows with coins ranging from the fifth to the early seventh century (*BASOR* 191, 17). A fourth rim, **380**, comes from a disturbed level north of BE-N (formerly BNH, *BASOR* 187, 58); with it came the lower part of a cylindrical bottle of fabric 2 (cf. **488**) and the rim of a bowl of the same fabric having a folded edge.

The last rim (**381**) was found in the area west of the Gymnasium (*BASOR* 191, 38) together with bottle necks (**496** and **559**) and ring bases (cf. **401–444**), all most likely datable to the sixth century.

374 *Pls. 12, 24.* G59.58:1804a. Salver with high foot with plain edge turned downward, short stem, wide shallow bowl with plain rim. Fabric 2, thin; mended.
H. 4.5, diam. foot 4.1, rim 12.5.
HoB area 5, niche in N wall. Found with **320** and **472**.
AJA 66, 11, no. 12, pls. 7:15, 9:12.

375 *Pls. 12, 24.* Foot (of salver?) with plain edge turned downward, short concave stem. Fabric 1.
P.H. 3.5, diam. foot 5.5.
BS E 16, 1963, to *96.40.

376 *Pl. 24.* Bell-shaped foot (of salver?) with plain edge turned downward. Yellow green tint (close to fabric 2); bubbly, irregular.
P.H. 3, diam. foot 5.
Pa-W 1967 E36.80-E43.80/N95.70-N104 *99.06-98.36.

---

62. Puttrich-Reignard, 20ff. A footed bowl of similar dimensions but datable to the 9th C. was found in Fustat: Pinder-Wilson and Scanlon (supra n. 15) 17, figs. 1–2.
63. Isings, no. 97a.
64. Cf. for example, Barag, "Glass Vessels," pls. 17, 30; similar bowls from Jeleme (northern Israel), 2nd half 4th C.; Saldern, *Slg. Hentrich,* nos. 157–159. For Egyptian examples cf. Edgar, pl. 2, no. 32.444; Harden, *Karanis,* 47ff., nos. 1ff.

Plain, Flaring Rims with Large Diameters

377 *Pl. 24.* One quarter of rim. Fabric 1.
Max. dim. 5.5, est. diam. 13, Th. to 0.1.
BS E 13, 1962 E76-E79/S2.5-S3.5 *96.40-96.10.

378 *Pl. 24.* One third of rim. Fabric 2.
Max. dim. 9.2, est. diam. 13, Th. to 0.1.
BS E 1, 1967 E10-E12.50/S0-S5 *97.25-97.00.

379 One sixth of rim. Fabric 1; decolorized, dirty fire scum.
Max. dim. 7, est. diam. 15, Th. to 0.1.
BS E 1, 1967 E7-E9/S0-S4 *98.75-98.00.

380 One quarter of rim, thickened edge. Fabric 1; decolorized, dirty fire scum.
Max. dim. 8, est. diam. 13, Th. to 0.15.
BE-N 1966 E32.7-E36.30/N96.7-N97.5 to *97.65.

381 *Pl. 24.* One sixth of rim. Fabric 1; eggshell.
Est. diam. 14.
W of B-W 1967 trench C W100-W102.70/N12.20-N14.20 *96.80-96.25. Found with **496** and **559**.

FOOTED BOWLS

A fragmentary, shallow footed bowl **382**, appears to be the only representative of this group found at Sardis. It has a bell-shaped foot, a short stem, and spreading walls curving upward (perhaps to a plain rim).[65] The complete vessel is closely related to salvers. Certainly more fragments of this bowl type have remained unrecognized. However, since feet identical to **382** could not be identified among the Sardis fragments (for very similar feet cf. **300-373**) it seems likely that this vessel shape was not common in the city. Discovered in one of the Byzantine Shops, its shape, the aquamarine color (fabric 1), and the mold-blown, ribbed vessel fragments (**651**) found with it seem to be proof of its Early Byzantine date.

The rim section of a large bowl or salver with an estimated diameter of about 30 cm. (**383**), having an upturned edge, may be a larger version of the same type. Unfortunately this piece is also not dated by archaeological context since it comes from a top level at HoB, but it was found with an Early Byzantine lamp base of type 3 (cf. **274-286**), (*BASOR* 170, 4ff.).

382 *Pls. 13, 24.* Irregular, bell-shaped foot, irregular concave stem, lower portion of bowl with wall almost horizontal then turning upward. Fabric 1, eggshell; dark spots.
P.H. 5, diam. at bowl fracture 6.6, original diam. certainly larger.
BS 1963, exact findspot unknown.

383 *Pl. 24.* One eighth of rim of large bowl (or dish) with upturned rim and probably shallow bottom (with foot?). Fabric 3, iridescent; scaling brown scum.
Est. diam. ca. 30.
HoB 1962 W30/S95 to *100.40.

MISCELLANEOUS STEMS

A group of miscellaneous, very fragmentary stems cannot be assigned to types; they appear to have affinities to goblets, footed beakers, and lamps. Their shape is biconical, apparently representing the lower half of the vessel bowl and the upper part of the foot of a goblet or beaker with conical foot and bowl similar to the attempted reconstruction of a vessel from Dura Europos.[66] They could also have formed part of lamps of the type represented by one nearly complete example and many fragments found at Jerash.[67]

The stems are either of pale green (fabric 2) or aquamarine (fabric 1) glass. With the exception of one piece that may have belonged to an entirely different vessel, **386**, all come from B. Two (**384** and **385**) were found together at Pa and are datable prior to the destruction of A.D. 616 (*BASOR* 191, 33); near them was found a salver foot, **376**, goblet stems, and bottle necks with spiral thread, **609**. In contrast to the other fragments, the join of bowl and foot of **385** is articulated by a knob. A third fragment, **386**, came from outside BS E 14 (*BASOR* 170, 49) with the bottom of a cup with ring base. This shop was particularly rich in glass finds, including the remains of goblets and lamps of various types, all belonging to the fifth and sixth centuries. **387**, was found in BE-A and belongs to a fifth to early seventh century level (*BASOR* 187, 12-14). Other glass finds in this area comprise plain, spreading vessel rims (of goblets or cups?), small handles (of lamps type 2?), a ring base (cf. **401-444**), and tesserae.

A fragment slightly more squat than the others (**388**) and showing the small portion of the faintly convex vessel wall, perhaps of a bowl, was found at

65. This bowl type appears to have been almost identical to a vessel form from Jerash preserved only in a very fragmentary state, datable to the 4th-5th C.; Baur-Kraeling, 525, no. 21, fig. 20 (here *Pl. 23*). Bowls from Jerash in the collection of the Ashmolean Museum, Oxford, have feet with a diam. of 4.5-5.8 cm. and a H. of ca. 3.5 cm. Cf. also Barag, "Glass Vessels," pl. 28. For related material from Egypt cf. Harden, *Karanis,* nos. 355ff., pl. 15.
66. Clairmont, nos. 744-745.
67. Crowfoot-Harden, no. 50; according to the authors of this study fragments "may equally have belonged to handleless drinking cups."

BE-N at a level overlaying the fifth to early seventh century strata; other fragments found with it include bases of cylindrical bottles, the rim section of a bowl with inturned edge, various plain bowl rims, goblet stems of type 1d, necks of bottles with spiral thread (cf. **607–621**), and the section of a lamp of type 3, all vessels common among Early Byzantine glass at Sardis. **390** no doubt belongs within this group but is not stratified.

384 *Pl. 24.* Double conical portion of upper foot and lower bowl of vessel. Clear with green tint (fabric 2?); iridescent, black spots.
P.H. 3, diam. at constriction 1.7.
Pa-W 1967 E36.80-E43.80/N95.70-N104 *99.86-98.36.

385 *Pl. 24.* Knobbed stem and egg-shaped lower portion of bowl. Greenish; brown scum, iridescent.
P.H. 2.8, diam. knob 1.9.
Pa-W 1967 E36.80/N95.70-N104 *99.86-98.36.

386 *Pl. 24.* Double conical portion of upper foot and lower bowl of a vessel. Fabric 1.
P.H. 5.3, diam. at constriction 1.2, at bowl fracture 4.8.
RTE 1962 E80-E84/S4-S10, fill over sidewalk in front of BS E 14, to *93.70.

387 Conical, waisted, rough pontil mark. Fabric 1; white spots.
P.H. 4, diam. at lower fracture 5.
BE-A 1966 E6-E8/N9-N9.9 *97.20-96.85.

388 Waisted stem. Green tint(?); dirty fire scum.
P.H. 2.5, diam. at constriction 1.7.
BE-N 1966 E32.7-E35.2/N99.5-N100 *97.95-97.45.

389 *Pl. 24.* Bell-shaped upper part of foot, constricted stem with funnel-like lower bowl. Yellow green tint.
P.H. 3, diam. at lower fracture 4.8.
PN 1964 W257-W261/S338-S341 *88.40-88.05.
This stem may be of later date, i.e. Islamic.

390 Waisted stem. Fabric 3.
P.H. 2, diam. at constriction 1.7.
HoB 1963 W15/S120 to *101.30.

SHALLOW BOWLS WITH WAISTED RIM

The large section of a bowl of fabric 1 glass (**391**) was found in the upper level of MTE (*BASOR* 177, 14–16). The wall curves upward to a waisted section below the flaring rim, and it is unlikely to have had a foot. It is the most complete bowl recorded at Sardis; however, a small number of rims and wall sections may have formed parts of such vessels.[68] Unfortu-

nately, **391** is not stratigraphically dated. The general area and level from which it came formed part of a dump that contained fragmentary early Roman as well as Early Byzantine material, the latter including remains of goblets and cups with ring bases (cf. **401–444**). Fabric and vessel type suggest an Early Byzantine date.

The lower portion of a bowl with waisted wall (**392**) is particularly interesting because it is made of turquoise blue glass. Found with part of a tube-shaped bottle (**556**), it certainly belongs to the fifth century (*BASOR* 174, 30ff.).

The rim section of a bowl (**393**) with profile similar to that of **391**, is made of fabric 1 glass and comes from the same unstratified level at MTE. Another rim fragment (**394**), having an edge bent horizontally, also appears to belong to this group. It is dated by coins and other evidence to the fifth to sixth century (*BASOR* 187, 54–58); a few sherds found nearby include plain vessel rims and the stem of an unidentifiable vessel (goblet?).

391 *Pls. 13, 24.* One third of body preserved. Apparently flat base, vertical wall with rim bent outward. Fabric 1; iridescent, brown flaking scum.
H. 3.2, est. diam. rim 7.
MTE 1964 E60-E65/S149-S153 *110.00-109.50.

392 *Pl. 24.* Flat base, wall slightly spreading. Turquoise-blue; white scum.
P.H. 1.2, est. diam. 4.
Syn MH 1963 E55-56/N3-N3.5 *96.50-96.00. Found with **556**.

393 *Pl. 24.* Vertical wall with out-turned rim. Fabric 1; brown scum.
P.H. 4.3, est. diam. 10, Th. 1.
MTE 1964 E60-E65/S149-S153 *110.00-109.50.

394 *Pl. 24.* One eighth of rim bent outward to horizontal position. Fabric 1; frosted white scum.
Max. dim. 3.5, est. diam. ca. 10, Th. to 0.1.
BE-CC 1966 E32.70-E35/N101-N102.50 to *97.20.

HIGH BOWLS

Fragments of a deep, slightly conical bowl with plain rim, **395**, were found in BS E 15. A vessel such

---

68. Waisted bowls of this type appear to be closely related to late Roman Imperial bowls generally supported by ring bases which were found in Syria (Palestine) and Egypt. Cf. Barag, "Glass Vessels," pls. 15ff., 31ff.; Harden, *Karanis*, 96ff., pls. 14–15 (almost always with folded rim); Edgar, pl. 2.

as this could have had a concave base with or without base ring or a foot with stem.[69] Among the other finds from this shop are bases of bowls(?) and handles, perhaps from lamps of type 2 (*BASOR* 174, 45; cf. also *BASOR* 170, 51).

The section of a slightly smaller bowl with faintly convex walls and thickened rim, **396**, has no provenance but seems to belong to this group; its fabric 3 glass is characteristic of Early Byzantine glass, and its shape also seems to link it with bowl forms of the same date.

395 *Pls. 13, 24.* G63.21:5903. About one fifth of vessel preserved, slightly conical wall; the vessel may have had a stem or a base ring. Fabric 2.
P.H. 7, est. diam. rim 10, Th. 0.1.
BS E 15, E87–E92/S1–S3 *94.10, foundation drain.

396 Slightly convex vertically oriented wall with rim thickened on ext. Fabric 3, iridescent, frosted, dark scaling scum.
P.H. 7.2, est. diam. 7, Th. 1.
Chance find 1962.

FOLDED FEET OF (HIGH?) VESSELS

Sloping feet with folded edge, apparently supporting a vessel, the bottom portion of which seems to have been cylindrical or almost cylindrical, may be related to the group of folded ring bases (cf. **449–462**). However, as the dating of all finds in this group is uncertain, the objects listed here may also be late Roman. The few pieces recorded are too fragmentary to suggest the original vessel shape or shapes with any certainty. Two have diameters of 6 cm. while the third is double in size.[70]

The large foot, **398**, and **397** come from east of Syn Fc. They can be associated with the construction of the "porch" in the third century as well as with later phases, extending to the sixth century (*BASOR* 177, 17–19); no exact dating can be provided. However, in the general area were found a large number of Early Byzantine fragments, including the remains of goblets, lamps, ring bases (cf. **401–444**). The third piece (**399**) is similar to **397** and comes from an unstratified level at AcT.

397 *Pl. 24.* G64.7:6519. Sloping folded foot, lower part of cylindrical vessel. Fabric 2; iridescent, frosted, brown scum.
P.H. 2.1, diam. 6.
E of Syn E120–E130/N5–N8 *97.00.

398 *Pl. 24.* Foot of large heavy vessel, similar to **397**. Fabric 3.

Est. diam. foot 12.0, Th. 0.2–0.3.
E of Syn 1964 E126–E129/N7–N10 *97.00-96.50.

399 Similar to **397**. Fabric 4.
Diam. 5.8.
AcT trench C 1960 D-B/5–8, fill, ca. *401.25.

400 *Pl. 24.* G73.1A:8233. Fabric 3, pale.
Diam. 5.0; est. diam. of vessel ca. 11.0.
PN/EA W208/S364, in bedding of earlier church floor, *89.56. Found with **236** and **246**.

RING BASES (OF BOWLS?)

This group includes numerous identical ring bases with which no body fragment can be identified. Occasionally rim sections of the same fabric were found with them, but it has proved impossible to link a rim with a base. They consist of a sloping base ring, a concave bottom and the lower section of a vessel with walls gently curving upward, almost invariably made of bottle olive (fabric 3), rarely of bottle green (fabric 4) and pale green (fabric 2), and almost never of the otherwise popular aquamarine (fabric 1) glass. All pieces recorded show similar weathering: dirty scum which, when it is scraped off or has fallen off, leaves a surface that displays a slight iridescent sheen.

By analogy to foot constructions of fourth to fifth century bowls from Karanis,[71] most of the bases of this type from Sardis could have belonged to a bowl or cup the lower part of which had a segmental cross section while the rim was flaring or slightly conical.[72] In a few cases the preserved wall above the foot ring

69. For a pattern-molded goblet with a bowl having a similar profile cf. Barag, "Glass Vessels," pl. 10; for deep, slightly conical bowls with or without base ring, datable to the late Roman Imperial period cf. ibid., pls. 15, 18, 31 et passim. For early 8th C. bowls with slightly convex walls and concave bases cf. K. Brisch, "Das omayyadische Schloss in Usais (II)," *MittKairo* 20 (1965) 173, fig. 42.

70. For a foot from Jerash similar to those from Sardis although without folded edge cf. Baur-Kraeling, 526, no. 27, fig. 21. Cf. also Barag, "Glass Vessels," pls. 26 ("11:7"), 33. A foot of similar profile which probably belonged to a vessel of the type represented by the material from Sardis, cf. Langdon and Harden (supra n. 50) fig. 5:17 (5th–6th C.).

71. Harden, *Karanis,* nos. 221ff., pl. 14. For beakers from this site having a very similar base construction cf. ibid., no. 362:4, pl. 15 (4th–5th C.). For predominantly late Roman Imperial bowls from Palestine with straight or slightly convex wall cf. Barag," Glass Vessels," pls. 15ff.

72. Cf. the rims found with **418**, **419**. Late Roman Imperial variants of footed bowls with a segmental lower body and feet of a profile very similar to that of the bases discussed here were found at Jeleme: Goldstein, pls. 21ff. (2nd half 4th C.).

extends to a height of about 3 cm.; here the wall section appears to have belonged to a vessel with an almost cylindrical lower half, possibly a bottle. But even then the upper vessel section could have increased in diameter to form a bowl.

Two plain rims having an estimated diameter of 10 cm. seem to stem from large vessels, perhaps bowls. They were found with **418**. Below the edge of one of these sherds (of fabric 3) the wall continues almost perpendicularly which might indicate that it formed part of a bottle, and not a bowl with outsplayed rim. Similarly, **419** was also found with two plain rims of fabric 3 glass that seem to have been part of straight-sided, spreading necks with estimated diameters of 9 cm. These rims, however, are not necessarily connected with the base rings.

The diameter of the base rings ranges from 4.0 to 6.4 cm. The estimated diameter of rims found with the base rings, made of fabric 3 glass and apparently from the same vessel, varies from 6 to 10 cm. In general the bases are fairly well made although they rarely approach the quality of the best goblets of type 1 (cf. **300–322**). They are sometimes irregular and less carefully finished.

One base, **401**, was found in BS E 4 as were other typically Early Byzantine pieces: goblet stems (including a multiknobbed stem, **369**), lamps, window panes, threaded bottle necks, etc. Other bases come from BS W 3, **402–404** (*BASOR* 154, 16–18) and BS W 10, **405**.

A large number of base rings (**406–415**) were discovered east of Syn (*BASOR* 174, 47; 182, 42ff.) where masses of fragments of lamps of various types, of goblets, and of a beaker came to light, all datable to the fifth and sixth centuries. Among the most significant vessel fragments found in the same area as the base rings were: the conical neck of a bottle (**505**) and two rim fragments of which one, with a diam. of 7 cm., comes from a conical vessel neck while the other, with folded edge (diam. 8 cm.), was part of a conical and spreading neck (found with **414**).

Portions of at least four base rings (**416–419**), come from BE-B and belong to the fifth to sixth century (*BASOR* 187, 16, there designated unit B′); lamp handles and window panes were found in the same room. **417** was discovered near the base of a large dish, **465**.

The many base rings found at HoB are less well dated than the finds from B and Syn. The finds from 1961 (**420–423**; *BASOR* 166, 5), 1962 (**424–429**; *BASOR* 166, 5ff.), and 1963 (**430–433, 435**; *BASOR* 174, 6–8) were found in disturbed levels overlaying Hellenistic occupation.

432 came with cullet; **428** was discovered near an early or mid-Imperial cup base similar to vessels listed under **185–187**, and window glass; **435** was found adjacent to a biconical stem section of an apparently large goblet. **436** was found in mixed fill with a straight, high base ring (not unlike **192**) and a plain, spreading rim with an estimated diameter of 7 cm., both made of fabric 3 glass.

In 1958, at least two base rings were found in Building L; **438** is from a level that may be earlier than the late Roman-Early Byzantine floor while **439** comes from just above this floor (*BASOR* 154, 10; cf. Hanfmann, *JGS,* 52; for the dating of the building, *Sardis* R1 [1975] 111, 114).

A few similar bases may complete this survey. One (**441**), identical in cross section to **442**—which is not securely dated—shows a greater height than the average base ring. It was found in BE-B (*BASOR* 187, 10ff.). **443**, of green tinted glass (close to fabric 2), comes from BE-N and is safely dated to the period antedating the destruction of A.D. 616 (*BASOR* 187, 57); in the same room were found fragments of salvers, cylindrical bottles, and probably handled lamps.

Unless noted otherwise the following ring bases are of fabric 3.

401 *Pl. 24.* Fabric 4.
Diam. 5.2.
BS E 4, 1967 E33.60/S0-S5 *98.00-97.00.

402 G58.22:408.
Diam. 5.
BS W 3, above level II.

403 G58.23:409.
Diam. 4.
BS W 3, above level II.

404 G58.27:387b. Fabric 4.
Diam. 5.
BS W 3, level II, floor 1.

405 Greenish-aquamarine (fabrics 1–2).
Diam. 4.4.
BS W 10, 1959 W52-W54/S2.50-S4.40 *96.50-95.50.

406 Diam. 5.
E of Syn 1965 E128-E130/N8-N12 *97.00-95.00.

407 Diam. 4.5.
E of Syn E125-E126.5/N1-S2 *97.75-96.80.

408 E of Syn E125-E128/N6 *97.80-97.00.

409 Greenish-olive (fabrics 3–4).
Diam. 5.
E of Syn 1963 E124-E126/N4-N9 *96.75-96.00.

410   Three bases.
Diam. 3.7–4.8.
Syn Fc 1965 E112.38-E122.50/N11-N15 *96.50-96.25.

411   Diam. 4.8.
E of Syn 1965 E124.47-E125/N11.5-N14 *96.75-96.25.

412   Diam. ca. 6.4.
E of Syn 1965 E124.47-E125/N11.5-N14 *96.75-96.25.

413   Two bases.
Diam. 4.7.
Syn Porch 1963 E116.8-E118/N4-N6 *96.75-96.25.

414   *Pl. 13*. Four bases. Three are fabric 4 and one fabric 3.
Diam. 4.5–4.8.
E of Syn 1964 E126-E129/N7-N10 *97.00-96.50. Found with rim fragment of a conical vessel neck (diam. 8 cm.).

415   Two bases. Fabrics 3 and 4.
E of Syn 1965 E120.5-E125/N11-N15 *96.50-96.25.

416   Portions of ca. 5 bases. Fabrics 3 and 4.
Diam. 4.5–5.
BE-B 1966 E18-E20/N7-N9 *96.80-96.40.

417   Diam. 4.8.
BE-B 1966 E16.30-E17.80/N3-N4 *96.70-96.35.

418   Four bases.
Diam. 5–5.2.
BE-B 1966 E17.50-E19/N5-N8 *96.80-96.60.

419   Diam. 5.2.
BE-B 1966 E20-E21/N4-N6 *96.80-96.40.

420   Diam. 5.3.
HoB 1961 W15/S95-S105 to *99.60.

421   Diam. 4.2.
HoB 1961 W5/S90-S95 to *101.50-100.50.

422   G61.9:3268.
Diam. 5.5.
HoB W10/S90 *100.30.

423   Diam. 4.7.
HoB 1961 E0/S90-S95 to *100.50.

424   Diam. 5.2.
HoB 1962 E5-E10/S115 to *101.40.

425   HoB 1962 E0/S105 to *98.30, all gravel.

426   Diam. 4.2.
HoB 1962 W20/S95 *101.50-101.00, fill.

427   Diam. 5.3.
HoB 1962 E25-E30/S85, clearing N wall of cistern, *100.00.

428   Three bases.
Diam. 4.5–5.2.
HoB 1962 E20/S90, area W of cistern, to *99.15.

429   Thirteen bases. Six are fabric 3; six are fabric 4; and one is fabric 2.
HoB, Lydian Trench 1962, upper fill.

430   Diam. 5.3.
HoB 1963 W13-W15/S120-S125 *102.80-102.50.

431   Diam. ca. 5.4.
HoB 1963 W3-W20/S120-S130 to *102.00.

432   Green-olive, fabrics 3–4.
Diam. 4.8.
HoB 1963 W30/S110 to *101.00.

433   Diam. 4.8.
HoB 1963 W10-W12/S110-S112 *99.50-99.30.

434   Diam. 4.1.
E of BS E 19, 1963 E122.5-E123/N5-N6 *97.30-96.75. Found with **505**.

435   Three bases. Two are fabric 4 and one is fabric 3.
Diam. 4.4–4.8.
HoB 1963 W15/S120 to *101.30.

436   Diam. 5.3.
HoB 1965, upper mixed fill, *102.40-100.50.

437   Diam. 5.3.
HoB 1962 E5-E10/S113-S126 *101.00.

438   G58.71:866.
Diam. 4.5.
L room B, surface to *98.60.
Hanfmann, *JGS*, 52-53; *Sardis* R1 (1975) 111.

439   G58.77:922. Fabric 2.
Diam. 5.2.
L room E, surface to *100.50.
Hanfmann, *JGS*, 52-53; *Sardis* R1 (1975) 114, listed as early Imperial.

440   PN 1963 W230-W234/S347-S350 *87.65-87.25.

441   Fabric 4.
Diam. 5.5.
BE-B 1966 E18.8-E19.6/N0.9-N2 *96.60-96.20.

442   *Pl.24*. Diam. 4.9.
HoB 1963 E0-W8/S120-S125 to *101.00.

443 *Pl. 24.* Fabric 2.
Est. diam. 6.
BE-N 1966 E26.50-E32/N96-N98 to *96.60, marble floor.

444 *Pl. 24.*G63.19:5829. Fabric 1. This base ring may belong to the group listed here.
Diam. 5.8.
E of BS E 19 E116-E119/S2-S4 *97.50-97.00.

FLAT BASES (OF CUPS?)

A series of bases has not yet been associated with any specific vessel type. They are flat while the vessel wall is convex, i.e. the lower portion of the bowl has the cross section of a bowl or cup. The diameter of three recorded specimens is about 6 cm. while the fourth piece is much larger. Two are of fabric 1, two of fabric 2.

445 comes from BS W 7 and may be datable to the sixth century (*BASOR* 157, 32). **446** was found in BS E 16; this shop revealed much Early Byzantine glass, including sections of lamps, bottles, and threaded bottle necks, as well as a salver foot, **375**. The other two pieces are not dated through archaeological context. One, **447**, comes from PN at a level containing Byzantine as well as late Islamic material (*BASOR* 177, 3); the other, **448**, is a chance find.

445 *Pls. 13, 24.* G59.22:1340. Irregular, sloping foot with convex wall of lower body. Probably fabric 2.
Diam. 4.7.
BS W 7 W22.00-W23.58/S2 *97.70-97.00.
AJA 66, 11, no. 16, pl. 10:16.

446 Irregular, top of base ring slightly sloping; must have belonged to large vessel. Fabric 2.
Diam. 8.8.
BS E 16, 1963 ca. E92-E98/S0-S4 *97.00-96.40.

447 *Pl. 24.* Fabric 1.
Diam. 5.9.
PN 1964 W255-W256/S339-S340 *88.35-88.20.

448 Fabric 1.
Diam. 5.9.
Chance find.

FOLDED RING BASES (OF CUPS OR BOWLS?)

These bases are similar to the ring bases (**401–444**) but have a folded edge and a vessel wall that usually spreads almost horizontally from the base, to curve gently upward. This profile could belong to a footed

bowl with perpendicular or flaring rim[73] but, the fragments preserved are too small to allow a reconstruction of the original vessel.

The bases are made predominantly of aquamarine (fabric 1), more rarely, of pale green (fabric 2) glass, and in at least one case, of bottle olive glass (fabric 3). Diameters vary greatly, from ca. 3.5 to 11 cm., with a median diameter of 5 to 7 cm. which seems to indicate that the larger bases come from vessels of a form different from that of the smaller. This seems to be borne out by the fact that the profiles show certain variations.

The undersides are usually convex. If a concave bottom with a folded base ring occurs it may mean that it was part of a different vessel type or the vessel was fashioned differently, receiving a concave pushed-up bottom.

Almost all of the bases were found in the Synagogue and the Byzantine Shops. **449**, **450**, and **451**, all found in 1965, are dated about the mid-fifth to the early seventh century (*BASOR* 182, 34ff.). Two from Syn Fc, **452** and **453** are certainly after the mid-fourth century (*BASOR* 170, 46; 177, 21); the first was with window fragments and handles, probably belonging to lamps of type 2. Much glass was found in the mixed fill of BS W 14 (*BASOR* 186, 29ff.), including a ring base (**455**), window panes, goblet feet, lamps of type 4 etc., all belonging to the Early Byzantine period (the fill contained mixed ware from Lydian to Byzantine times and coins of the fourth and fifth centuries). Another base (**456**) from BS E 15 is no doubt contemporary with the Byzantine glass discovered in the neighboring shops (*BASOR* 170, 51). Two more bases, **457** and **458**, are of the same date although they were found under the sidewalk in front of BS E 14 and in BS W 13.

A folded base of slightly different form (**462**) may have been part of another vessel type; as it is unstratified, it could conceivably be late Roman Imperial.

449 Fabric 1.
Est. diam. 6.
Syn 1965 E71-E73/N21-N24 *95.00-93.50.

450 Fabric 1.
Est. diam. 10.
Syn 1965 E90-E93/N1.5-N3 *95.60-95.50.

451 *Pl. 25.* G65.1:6786. Complete. Fabric 1.
Diam. 6.5, Th. 0.05.
Syn SE pit E90/N2-N5 *94.20.

---

73. Cf. the type in Harden, *Karanis,* pl.14:266.

452 Fabric 3, crudely made.
Est. diam. 8.
Syn Fc 1962 E115.5-E116.5/S0-S2 *97.50-96.50.

453 Fabric 2.
Diam. 3.6.
Syn Fc 1964 E129-E130/N6-N9 *96.50-96.00.

454 Portions of 2 base rings. Fabric 1, very thick. Heavily fractured.
Diam. 7, Th. 0.2–0.5.
BS E 13, 1962 E75-E76/S2-S9 *96.70-96.20.

455 *Pl. 25*. One half of base ring and lower vessel with convex wall. Fabric 2.
P.H. 2.7, diam. 4.3, Th. 0.2.
BS W 14, 1966 W65.70-W68.50/S3.10-S4.80 *97.00-96.30.

456 *Pl. 25*. Fabric 1.
Diam. 4.
BS E 15, 1962, pit under shop, *94.50-94.00.

457 Similar to **451**. Well made.
Diam. 6.3.
RTE 1962 E81-E85/S4-S10, fill under sidewalk in front of E 14, to *93.70.

458 *Pl. 25*. One half of base ring. Fabric 1, well made.
Diam. 5.3.
BS W 13, 1959 W54-W57/S2-S4 *94.50-94.00.

459 White iridescent scum.
Diam. 3.8.
HoB 1963 W15/S110-S115 to *100.70.

460 Base ring with slightly convex wall. Fabric 1.
Est. diam. 13, Th. to eggshell.
MTE 1964 ca. E60-E65/S150-S155 to *112.20.

461 *Pl. 25*. Many bases.
Diam. 3.8–7, median diam. 4.8–5.8.
MTE 1964 E60-E65/S145-S150 to *110.00.

462 *Pl. 25*. Base ring with vertical lower part of vessel. Fabric 2.
Diam. 6.
AcT trench E 1960 fill.

LARGE DISHES

Lower portions of apparently shallow dishes with conical base rings have a close similarity to early and mid-Imperial ware. The base ring was probably not made together with the dish but was attached later. As no rim fragment has been identified that without question belongs to a large plate with base ring, it is not possible to offer a safe reconstruction of the entire vessel. Dishes with base rings such as those described here probably had walls rising gently to a plain rim like the generally plain-edged footed salvers and rims of large vessels that are found in Early Byzantine levels. The plain rim may have been bent downward. The profile of the lower part should be compared to Roman dishes of the third and fourth centuries which may have been the predecessors of the Byzantine vessels.[74]

An almost complete base ring and the fractured remains of the central portion of the bowl are preserved in **463**. As the diameter of the base is 18 cm., the entire dish must have been exceptionally large. It is made of yellow brown glass, a material related to fabric 3. The fragments of the dish come from the Packed Columns Area east of BS E 19 (*BASOR* 174, 45-47) where the latest coins are those of Justinian (A.D. 527–565, indexed under PCA in *Sardis* M1 [1971] 150). At this general location and level were found many glass vessel fragments attributable to Early Byzantine times as well as coins of the fifth and sixth centuries.

In the immediate neighborhood of this large base ring was found the rim section of a very large dish (**464**) of fabric 2 glass with down-turned edge; the diameter measures 44 cm. The size of **463** and **464** lead one to suppose that they may have come from vessels of very similar or identical shape, namely exceptionally large dishes (or shallow bowls?) with base ring and a flattened or down-turned rim. **465** comes from BE-B at a level that contained much Byzantine glass, datable to a period after ca. A.D. 400 but before the post-616 squatter occupation (*BASOR* 187, 16). Also discovered there were fragmentary ring bases (cf. **401–444**), lamps, small handles of lamps, cullet, and window panes.

Base ring (**466**) from below floor level in BS E 17 (*BASOR* 174, 45–48) appears to be earlier than the Early Byzantine glass from this shop. The profile resembles that of the other examples just mentioned, but its fabric and general appearance are close to that of the heavily iridescent Roman cups (cf. **185–187**). Pottery was also found at this level, apparently antedating the Byzantine material. It is possible that this base ring is of late Roman date, which would be an

---

74. Immediate predecessors of this type seem to be represented by shallow dishes with ring bases from Jeleme, 2nd half 4th C.; cf. Goldstein, pl. 23:5. Cf. also a perhaps late 4th C. shallow dish of similar profile with diamond-point engraved decoration: Avigad (supra n. 58) 229–230, fig. 12; Barag, "Glass Vessels," pl. 30. Cf. also Harden, *Karanis*, 49ff., pl. 11:4, 7.

indication that vessels of this type were popular throughout late Roman and Early Byzantine times.

Other bases with similar profiles are unstratified. One (**467**), from the upper mixed fill at HoB, came with a large quantity of glass fragments of mainly Early Byzantine date, including goblets, lamps, and ring bases (cf. **401–444**). Bases from HoB (**468–470**) are undated as well (*BASOR* 166, 5–6; 170, 13). With **471** were found (Byzantine?) window panes and rim fragments of probably late Roman cups with curved rims (cf. **200–203**).

463 *Pls. 13, 25.* Sloping base ring with fractured, almost horizontal wall section of central part of plate; no rim fragment preserved. Pale yellow-brown (bottle olive?); this fabric seems to be slightly different from the regular fabric 3 ware; mended.
Diam. base ring 18; Th. 0.20–0.25, at center 1.
PCA 1963 E116.5-E121/S1-S3 and E120-E124/S1-S3 *97.50-97.00.

464 G63.20:5880. About one fifth of rim of a large dish or shallow bowl, rim bent downward. Fabric 2.
Diam. 44; Th. 0.2–0.3, rim 0.5.
PCA E118-E121/S0.50-S2.50 *95.50-95.25.

465 *Pls. 13, 25.* Conical base of dish with fractured edge of lower body. Fabric 1.
P.H. 1.9, diam. base 5.6.
BE-B 1966 E16.30-E17.80/N3-N4 *96.70-96.35.

466 *Pl. 25.* Coil-like base ring (of bowl?). Fabric 2; heavy iridescence, milky scum.
Diam. 6.
BS E 17, 1963 E98/S1-S9 *94.70.

467 *Pl. 25.* One fourth of base ring. Fabric 1.
H. of ring 1.6, est. diam. 13.5.
HoB 1964 E60-E65/S150-S155 *109.00.

468 Part of base ring. Fabric 2.
Est. diam. 14.
HoB 1961 W5/S105, ramp, to *99.90.

469 Section of base ring. Fabric 2.
Est. diam. 9.
HoB 1962 E5-E10/S115 to *100.40, clearing pipe.

470 *Pls. 13, 25.* Section of base ring. Same fabric as **463**.
Est. diam. 14.
HoB 1965 *102.40-100.50, upper mixed fill. Found with **579**.

471 Base ring. Yellow green tint; heavy scaling iridescence.
Diam. 5.8.
Chance find 1959.

BEAKERS

Among the most important finds is a fairly well-preserved bell-shaped beaker (**472**), a shape which derives from plain beakers of mid- and late Roman Imperial times.[75] Close parallels to the Sardis finds are afforded by a few Early Sassanian beakers from Tell Mahuz in the Iraq Museum in Baghdad. According to the numismatic evidence all glass from Tell Mahuz should be datable to the late third and fourth century, a period preceding the most productive time in Sardis by one or two centuries.[76]

The Sardis beaker (**472**) was found with a salver

75. Isings, no. 106c and H. J. Eggers, *Der römische Import im freien Germanien* (Hamburg 1951) form 230. For many variants of the late Roman beaker cf. A. Benkö, *Üvegcorpus, Régészeti Füzetek* (Budapest 1962) II pls. 28ff., cf. esp. no. 12/é, pl. 31:7. Other 4th C. bell-shaped beakers preceding the Sardis type—often having a ring base—were found in Jeleme and Naharya (I am most grateful to Gladys Davidson Weinberg for showing me the material stored in Jerusalem). For beakers more distantly related to the finds from Sardis cf. W. Slomann, *Saetrangfunnet* (Norske Oldfunn IX; Oslo 1959) pl. 9:1 (4th–5th C.); from Carmona: *Guia del museo y necropolis romana de Carmona (Sevilla)* (Madrid 1969) pl. 16. Cf. also Weinberg (supra n. 49) 133, pl. 25, fig. 4C.

76. Many coins of Shapur II (A.D. 309–379): M. Negro Ponzi, "Sassanian Glassware from Tell Mahuz (North Mesopotamia)," *Mesopotamia* 3–4 (1968–69) 308–309, nos. 63–64 (not all of the parallel pieces cited by the author are comparable to the beakers from this site); idem, "A Group of Mesopotamian Glass Vessels of Sassanian Date," *London Congress,* 13, fig. 3 (the material is preserved in the Iraq Museum, Baghdad). Beakers identical to these Mesopotamian finds have turned up lately in the antiquities market; a beaker very similar to the Sardis beaker, 15 cm. H. has allegedly been found in Iran. For a beaker of the same form, the lower part of which shows a pinched pattern of "nip't diamond waies" (cf. **92–93**) see *JGS* 6 (1964) 159, no. 13. Finds from Jerash, including slightly waisted beakers with pushed-up base, may be a bit earlier than, or contemporary with, the beakers from Sardis: Crowfoot-Harden, pl. 28:4. Beakers (or lamps?) with three small handles, which have a cylindrical or slightly conical lower body and a cup-shaped upper portion, were found in Hira and Nineveh (the material is in part preserved in the Ashmolean Museum, Oxford). For general remarks on the finds from Nineveh see Harden, "Nineveh" (supra n. 61) 186. The upper portion of a beaker or lamp with slightly flaring rim and three handles, datable to the late 5th or 6th C.: Wiseman (supra n. 7) 105–106. For unstable bell beakers of the 5th and 6th C. found in the West cf. D. B. Harden, "Glass Vessels in Britain and Ireland, A.D. 400–1000," *Dark-Age Britain: Studies Presented to E. T. Leeds* (London 1956) 140–141, fig. 25.

(**374**) and the foot of a goblet (**320**) in a niche of unit 5 at HoB (*BASOR* 157, 24–26). The good condition of the beaker and salver is exceptional; moreover, both vessels appear to represent less common shapes. The house in which they were found was inhabited until the early seventh century; a coin of the sixth century and coins of Heraclius help to date these pieces to the sixth or early seventh century. No doubt more beakers are represented among the numerous wall fragments which could not be connected with any known vessel form.

A badly deformed vessel softened by fire (**473**) appears to have been a beaker; its height is similar to that of **472**. It comes from an area east of Syn Fc rich in glass finds and coins ranging from Constantine I to Justinian (*BASOR* 174, 47; *Sardis* M1 [1971] s.v. Find-spot Index, Syn-Porch, 150), thus placing the piece in the fifth or sixth century.

The portion of a cylindrical vessel, **474**, found under BS E 15 was certainly part of a beaker. Although it came from a relatively low level in the fill underneath the floor, it could date to the early(?) fifth century (*BASOR* 170, 51). Of later date, probably the early seventh century, are a few rim fragments which may be remnants of beakers (**475**). Found with various base fragments of vessels (beakers?), they come from a comparatively high level overlaying the opus sectile floor of BE-N which is datable to the late fourth or early fifth century (*BASOR* 187, 55–57).

472 *Pls. 13, 25.* G59.42:1483. Bell-shaped with plain rim, slightly concave base. Fabric 1; mended, small sections missing.
H. 12.7, diam. rim 9.7.
HoB area 5, niche of N wall. Found with **320** and **374**.
BASOR 157, 25, fig. 11; *AJA* 66, 11, no.13, pl. 7, fig. 16, pl. 10:13; Hanfmann, *Letters,* 59, fig. 36.

473 *Pl. 13.* G63.14:5373. Deformed; the present state of the piece seems to indicate that it had been a beaker. Fabric 2 (?); black fire scum, fire-softened.
P.H. 13.
E of Syn E116.7-E117.8/NS0 *98.00-97.75.
BASOR 174, 47.

474 *Pl. 25.* Cylindrical lower body curving to concave base. Green tint (fabric 2?).
P.H. 7.3, est. diam. ca. 8.
BS E 15, 1962, fill under shop, *95.30-95.00.

475 Two concave bases of which one curves to cylindrical body. Three rim fragments one of which shows a large portion of a cylindrical body. Probably fabric 2, thin; dirty fire scum.

Est. diam. bases 4.7 and 4; est. diam. rims 7, 7.4, and 7.7; P.H. of largest rim fragment 6.
BE-N 1966 E32.7-E35.2/N99.5-N100 *97.95-97.45.

BOTTLES

The bottle in all its forms and variations is the most common glass vessel in Roman and post-Roman times. In Sardis a few other categories, such as lamps and goblets, seem to have been made in similar quantities, but the bottle, in a variety of different shapes, is the most frequently found vessel.

With the exclusion of miniature vessels, only very few remnants of bottles are large enough to serve as a guide for the reconstruction of a complete vessel. A number of clearly recognizable shapes are listed, followed by variously shaped necks and bases of bottles of more or less uncertain shape. The uncertainty in classifying specific types with the help of small fragments mars the catalogue listing: each fragment may, or may not, be representative of a specific bottle type; each neck or base could be grouped with one or another type.

Type 1: Bottles with Ovoid, Oval, or Spherical Body

A fairly large vessel often with concave base, ovoid body and tall, usually slightly conical neck is a frequent type at Sardis. Variants include an oval or spherical body and a neck having a funnel-like or cylindrical upper part of greater diameter than the lower. This type was popular in the East in late and post-Roman times.[77]

77. The full range of bottle shapes in late and post-Roman glass found in Syria (Palestine) is illustrated in Barag, "Glass Vessels." Oval and ovoid bottles from Jeleme some of which have spiral threads around the upper part of the neck (cf. **607–621**) are datable to the 2nd half 4th C., i.e. immediately preceding the Sardis material: Goldstein, pl. 20; P. N. Perrot, "The Excavation of Two Glass Factory Sites in Western Israel," *Bruxelles Congrès,* 258:3, fig. 3a. Cf. also Bagatti-Milik, 19, fig. 34:1–5. For Early Byzantine bottles of the types discussed here cf. for example: Barag, "Netiv," fig. 2:4 (mid-5th-early 7th C.). L. Y. Rahmani, "Mirror-Plaques from a Fifth-Century A.D. Tomb," *IEJ* 14 (1964) 52–53, fig. 2:2, pl. 15:B (from Kfar Dikhrin). Delougaz-Haines, pl. 59:2 (from Khirbet el-Kerak, ca. early 5th-early 7th C.). Barag, *Shavei Zion,* 65ff., nos. 1ff., fig. 16:1–3 (5th–6th C.; with ref. to additional material; here *Pl. 25*). Fragmentary bottles from Auja Hafir, preserved in the Ashmolean Museum, Oxford (cf. Harden, *Nessana,* 86ff.). Bottles of this general form have also been found in South Russia: from Konstantinovkaja, Kerch (Ashmolean Museum, Oxford). A bottle of allegedly 6th C. date with perhaps spherical body and long neck,

Unfortunately the only three fairly complete bottles with ovoid body are without provenance (**476–478**). However, their fabrics (1 and 2) and the similarity of their necks to securely dated bottle necks make it certain that they belong to the Early Byzantine period. **476** has a slightly conical neck that increases in diameter, funnel-like, at the lip which has a plain edge; the lower portion of the body is missing but the base was no doubt concave.[78] **477** has a straight conical neck with an irregular, infolded rim. The neck of the third example, **478**, with a portion of the base preserved, is cylindrical at the bottom and funnel-like at the top, derived from a neck construction in vogue in Syria especially in the fifth and sixth centuries.[79]

The height of the three bottles varies from about 13.5 to 16.5 cm. The necks usually represent about one half of the total height, or an average height of about 6 to 8 cm.

The neck of **479** is identical to that of **476**. It is fractured at the upper shoulder, showing the wall curving outward to form the body which most probably was ovoid. It was found inside the large drain along the west side of the west wall of MC which also contained Early Byzantine goblets, cylindrical bottles, windows, threaded necks (cf. **607–621**), and cullet (*BASOR* 177, 23).

Among the conical necks that might be associated with ovoid, oval-spherical, or even cylindrical (cf. **488**) bottles, two are made in fabric 1. One (**480**) comes from BE-E while the other (**481**) was found, together with the remains of goblets, part of a bowl lamp(?), and a pattern-molded bottle as well as other Early Byzantine glass, at PN at a level datable to the fifth century (*BASOR* 166, 16ff.). Another variation of a vessel neck is represented by the portion of a bottle found at BE-H (**482**) at a level (*97.40–97.30) that suggests an Early Byzantine date (*BASOR* 177, 25). Although similar to funnel necks (**476**) the cylindrical lower part increases in diameter at a lower point. **482** was found with a slender, conical bottle neck (**489**) and other Early Byzantine glass, including upward-tapering bottle necks, concave bases, ampullae, and plain rims of tapering vessels. A similar bottle neck was found in BS E 16 with a foot possibly belonging to a salver (**375**).

An almost complete section of the lower half of a bottle with oval, almost spherical body, (**483**) seems to be dated fairly accurately as it comes from unit BE-C (*BASOR* 187, 19; referred to as C′). The vessel seems to belong to the period of remodeling of the room early in the fifth century. With this section were found two ring bases (cf. **401–444**) and two concave bases, perhaps of cylindrical bottles. Fragments of about eight bottles (**484**), "perhaps contained in a large jar" (identification tag) came to light with other Early Byzantine material in the northeastern corner of BS E 12 (*BASOR* 177, 20). The neck sections that were found with them are very similar to those of this group, i.e. they are either conical, or slender with diameter increasing toward the lip, or short and flaring toward the top. These necks range in height from 3.5 (short, flaring necks) to 9 cm.; the bodies of these bottles seem to have had an average height of 6 to 8 cm. Thus ovoid and oval-spherical bottles were very similar indeed as far as height and form of base and neck are concerned.

Among the fragments from neighboring areas rich in Early Byzantine glass are parts of the neck and shoulder of a bottle from BS E 16 (**485**) and the curved section of a probably oval or spherical bottle (**486**) from BS E 19. The area immediately to the east of BS E 19 also yielded fragments of constricted bottles, base rings of dishes (for example, **464**), vessel rims, a beaker (**473**) and coins, ranging from Constantine I to Justinian (*BASOR* 174, 47).

476 *Pls. 14, 26.* Upper part of body; ovoid, almost horizontal shoulder, slender neck increasing in diam. toward irregular, plain rim. Fabric 1, eggshell, bubbly.
P.H. 15.5, est. H. 16.5, diam. 7.5.
Chance find.

477 *Pls. 14, 26.* Neck and upper part of body (mended). Probably ovoid body with concave base, conical neck with irregular, infolded rim. Fabric 2, eggshell; iridescent, heavy dark scum.
P.H. 12.5; est. H. 13.5; diam. 6, of rim 3.
Chance find.

478 *Pls. 14, 26.* Almost spherical body with concave (pushed-up) base, cylindrical neck with high funnel mouth; mended, portions of body and neck missing. Pale fabric 1, eggshell, bubbly.

---

having a lower constriction and two applied threads: Crowfoot, *Samaria*, 418, fig. 99.

For a late stage in the development of this type, with neck tapering upward, cf. Brisch (supra n. 69) 173, fig. 39.

78. For bottles with spherical body and cylindrical neck the upper part of which is likewise cylindrical but of greater diam. cf. Barag, "Glass Vessels"; idem, "Netiv," fig. 2:5, pl. 27:4 (ca. mid-5th –early 7th C.).

79. Bottles of the same type: from Kfar Dikhrin, Rahmani (supra n. 77) 52–53, fig. 2:1, pl. 15:C. Crowfoot, *Samaria*, fig. 95. A bottle with ovoid body, narrow neck, and funnel mouth: Saldern, *Slg. Hentrich,* no. 193. For late 4th C. Western examples cf. Isings (supra n. 15) 262:4, fig. 5.

H. 14.; diam. 7, of rim 3.4.
Chance find.

479 *Pl. 26*. Neck and upper shoulder of bottle identical to
476. Clear with yellow tint, eggshell.
P.H. 8.
BE-E 1964 E11-E12.1/N42-N44, inside large drain along
wall of MC *96.50-96.00.

480 *Pl. 26*. Conical neck with slightly thickened rim, lower
neck spreading to body (ovoid, spherical or cylindrical).
Fabric 1.
P.H. 11.5, est. H. of bottle ca. 20, H. neck 10, diam. rim 4.
BE-E 1964 E11-E14/N32-N36 *97.00-96.50.

481 *Pl. 14*. G61.19:3771. Conical neck on probably ovoid
or slightly conical body. Pale fabric 1.
P.H. 10.3, diam. rim 4.9.
PN W265/S360 *88.45-88.33. Found with **490** and **646**.

482 Portion of neck with funnel mouth and shoulder on
probably spherical or cylindrical body. Fabric 1, thin.
P.H. 9; diam. shoulder ca. 6, of neck 2.
BE-H 1964 E12/N50, from hole below floor level, *97.40-
97.30. Found with **489**.

483 *Pl. 26*. Part of base and wall of bottle with concave
base and squat, almost spherical body. Fabric 1, eggshell;
heavy multilayered decomposition.
P.H. 6, est. diam. 8.
BE-C 1966 E31.9-E33.6/N13.09-N15.15 *97.50 down.

484 Fragments of ca. 8 spherical or near spherical bottles,
conical necks or cylindrical necks with funnel mouth. Fabrics 1 and 2, eggshell; decolorized.
P.H. bodies ca. 8, necks 3.5–9.
BS E 12, 1964 E74/S0-S1, next to back wall on floor, *96.50-
96.25.
BASOR 177, 20.

485 Cylindrical neck (rim missing) spreading to body with
rounded shoulder. Clear with green(?) tint; dirty fire scum.
P.H. 7, diam. neck 2.4.
BS E 16, 1963.

486 Curved section of bottle (?). Greenish with yellow tint;
iridescent, corroded.
Max. dim. 7, Th. 0.2.
BS E 19, 1963 E116.5-E121/S1-S3 *97.25-97.00.

487 *Pl. 26*. G61.22:3774. Portion of shoulder and tapering neck of bottle. Fabric 1; mended.
P.H. 4; diam. at lower fracture 6.8, at neck fracture 2.7; Th.
0.1–0.2.
PN W265/S360 *88.45-88.33. Found with **237** and **518**.

Type 2: Cylindrical Bottles

Only one fairly complete example of a cylindrical
bottle with concave base, slightly bulging shoulder
and conical neck was found at Sardis (**488**). However,
many of the concave bases and conical necks as well as
unknown quantities of small sherds from vessels similar to, or identical with, the type discussed here may
be diagnostic of the frequency of this form.[80]

The estimated height of **488**—a small section of
the central body is missing, upper and lower portions
do not join—is about 20 cm. It was found just south
of BS E 12 and according to its excavator, J. S. Crawford, probably belongs to the late sixth or early seventh level.

A few concave bases with a curve toward the lower
portion of an apparently cylindrical body could conceivably have belonged to this type, but because their
reconstruction is very tentative, they have only been
listed in the statistical tables. (Cf. esp. **524**; about six
come from BS E 13, others were excavated in the area
east of Syn Fc, in BS E 12, BE-B, and BE-N).

488 *Pls. 14, 26*. G68.3:7736. Neck, upper and lower portions (not joining) of bottle; section at lower body missing.
Cylindrical with concave base (no pontil mark), pronounced
shoulder, conical neck. Fabric 1.
Est. H. ca. 20; P.H. of upper portion and neck 14.3; H. neck
7.9, of lower portion 2.5; diam. base 7.
BS E 12 E71.50-E72.25/S4.50-S5.50 *96.00.

Type 3: Conical Necks and Type 4: Conical Necks Tapering Upward

Bottle necks which range from conical to cylindrical
to upward-tapering and are generally made of fabrics
1 and 2[81] were no doubt originally from bottles with
ovoid, oval, spherical, or cylindrical bodies (**476–**

---

80. Fragmentary bottles with slightly conical or cylindrical neck
of which at least the upper portion of the body is cylindrical were
found in Auja Hafir; they seem to be 5th–7th C. (preserved in the
Ashmolean Museum, Oxford; cf. also Harden, *Nessana*, 86ff.). Cylindrical bottles with conical neck: W. Martini, "Samos-Stadt:
Römische Thermen," *AA* (1975) no. 1:42, fig. 31, found with coins
of A.D. 600–620 and 650–670. For an earlier phase of this type,
having a short neck with a broad, folded rim cf. Barag, "Glass Vessels," pls. 9, 45 (here *Pl. 25*). Pattern-molded, ribbed bottles of cylindrical shape and with funnel mouth were found in Jeleme, 2nd
half 4th C.: Goldstein, pls. 21ff.
81. In addition to the bottles with conical neck listed in the
Barag corpus and in the foregoing notes cf. conical bottle necks
datable to the 6th C. or later, D. B. Harden, "The Glass Found at
Soba," *Sudan Antiquities Service. Occasional Papers* 3 (1955) 63, no.
13, fig. 37.

488). Those with conical wall tapering downward seem to have been most numerous. Over a hundred were found in the Syn and B area and more came to light in most of the Byzantine Shops, proof that these bottles were very common in Sardis.

A slender bottle neck (**489**) from BE-H was found with various fragments, including **482** and a similar but more conical neck of thick glass (similar to **515**), concave bases of bottles and ampullae, and plain rims, all certainly of Early Byzantine date.

Other conical or almost conical necks were found in places where much Byzantine material came to light: PN, within an Early Byzantine context (**490**); Pa-S (**491**); Syn Fc (**492** and **493**); BE-H (**494**), many in BS E 13 (**495**); W of B (**496**, found together with ring bases, cf. **401–444**, plain rims, and necks with folded edges; and BS W 13 (**497**).

A neck which tapers upward (**499**)—much rarer than those tapering downward—was also found in BS E 13.

489 Conical neck with thickened rim. Clear, green tint (fabric 2?).
P.H. 7, diam. 3.
BE-H 1964 E12/N50, in hole below floor level, *97.40-97.30. Found with **482**.

490 G61.20:3772. Conical neck and upper part of shoulder. Pale fabric 1.
P.H. 5.5, diam. rim 3.1.
PN W265/S360 *88.45-88.33. Found with **481** and **646**.

491 Conical neck and part of shoulder. Fabric 1.
P.H. 7, diam. bottom of neck 2.8.
Pa-S 1966 E43-E45/N20-N29 *96.50-96.00.

492 One-half plain rim and slightly conical neck. Fabric 1.
H. neck 3.8, diam. lower neck 1.6, est. diam. rim 2.8, Th. 0.15.
Syn Fc 1963 E101-E105/N1.20-N3 *97.50-97.00.

493 Sections of bottle necks: cylindrical and slightly conical, one with folded rim. Fabric 1.
Syn Fc 1963 E113-E115/N2-N6 *97.50-96.80.

494 Part of cylindrical (?) neck and shoulder of bottle. Fabric 2.
Max. diam. 4.
BE-H 1964 E10-E15/N75-N78 *97.00-96.50.

495 Bottle necks.
BS E 13, 1962 E75-E80/S0-S3.5 to *96.50 floor.

496 *Pl. 26*. Slightly conical neck spreading to shoulder. Greenish (fabric 2?); dull, white scum.
P.H. 7, diam. at lower neck 3.

W of B trench C 1967 W100-W102.70/N12.20-N14.20 *96.80-96.25. Found with **381** and **559**.

497 G59.62:2150. Conical neck curving to shoulder. Fabric 1.
P.H. 8, diam. rim 0.4.
BS W 13 W54-W57/S2.00/S4.40 *95.50-95.00.

498 Conical neck with sloping shoulder. Fabric 1.
P.H. 9.0, diam. rim 3.5.
BS 1969 W56-W58/N3-N7 *96.80.

499 Section of lower neck tapering upward. Fabric 2.
P.H. 9.2, diam. 3–4.5, Th. 0.15–0.2.
BS E 13, 1962 E75-E76/S2-S9 *96.70-96.20.

Type 5: Neck with Bowl-Shaped Orifice

Another type of neck is represented by two portions of upper necks of bottles (**500** and **501**) which consists of a cylindrical lower part and an upper portion which flares and then turns toward the rim in a bowl-like fashion.[82] The first was found in Building B but the exact findspot is unknown; the provenance of the other is unknown. Both may have belonged to cylindrical or globular bottles certainly of Early Byzantine date. A number of cylindrical necks with their upper portions missing—recorded here only statistically—may have belonged to this type.

500 Upper part of cylindrical bottle neck with funnel mouth and folded rim. Fabric 3, badly made.
Diam. rim 5.
B 1959 fill, *99.00-97.30.

501 *Pl. 26*. Neck slightly tapering to wide orifice. Fabric 1.
P.H. 6.5, diam. rim 4.8, diam. neck 2.8.
Chance find 1961.

Type 6: Wide Funnel Necks

A variant of the foregoing type is a short funnel neck, partly conical (i.e. straight-sided) and partly flaring (i.e. slightly waisted). Again the original bottle shape must remain unknown although it is likely that it was cylindrical.

A few of the necks come from BS. One, **502**, was found in BS E 13 (*BASOR* 170, 50). A more conical

---

82. This neck seems to represent a variant of the funnel-like upper portion of the neck discussed in **476–487**. Cf. also, for example, Barag, "Glass Vessels," pl. 42:20, illustrating a post-Roman bottle with a neck that is a cross between the funnel-shaped and bowl-like upper portion of the necks found in Sardis.

neck (**503**) from BS E 16, came together with devitrified, cullet-like lumps, tesserae, and a fractured rod
or stick (**676**; see *BASOR* 174, 45). **505** was found with
a ring base of the same fabric (**434**). A funnel neck
from LNH 1 (**506**), the fabric of which is similar to a
beaker lamp of type 4 (**290**), may be datable to the
fifth or sixth century (*BASOR* 187, 57–58).

502 *Pl. 26*. Funnel neck flaring to thickened rim and
spreading to rounded shoulder (of cylindrical or spherical
body). Fabric 1.
P.H. 6, diam. rim 3.5.
BS E 13, 1962 E75-E80/S0-S3.5 to *96.50 floor.

503 Conical neck curving to almost horizontal shoulder
(of cylindrical body?). Fabric 1.
P.H. 4.5, H. neck 3.8.
BS E 16, 1963, to *96.40. Found with **676**.

504 *Pl. 26*. Funnel neck curving to body; mended. Fabric
3, bubbly.
P.H. 3.5, diam. rim 4.8.
E of Syn 1963 E122.5-E123/N5-N6 *97.30-96.75.

505 Slightly irregular shape. Fabric 3.
E of BS E 19, 1963 E122.5-E123/N5-N6 *97.30-96.75.
Found with **434**.

506 *Pl. 26*. Funnel neck with infolded rim, curving to
shoulder. Clear with green tint; bubbly, black spots. Fabric
different from other Early Byzantine fabrics, relatively carelessly made.
P.H. 4.5, H. neck 3.5, diam. rim 3.7.
LNH 1, 1966 E40.40-E44.83/N103.26-N110.25 *96.50-
96.30.

507 Neck flaring to slightly in-turned rim. Fabric 2.
Diam. rim 5.
AcT 1962 E5/N26 *405.30.

508 *Pl. 26*. Upper portion of bottle. Oval (or cylindrical?) body, conical neck with folded rim. Clear.
P.H. 6.5; diam. rim 3.1.
BE-C 1973 E15-E17/N22-N23 *99.00-98.19. Found with
621.

Type 7: Globular Bottles with Wide Neck

Bottles with an ovoid almost spherical body and
wide, conical neck are represented by an almost com-
pletely preserved example and a few fragmentary
vessels and large sherds. This shape seems to have
also been made with added thread decoration (cf. **622
–632**).

The complete bottle, **509**, comes from a grave
(72.2) below the floor of the sunken area in the south

aisle of Church E at PN (*BASOR* 215, 33–37) and
should probably be dated to the very late fourth or
fifth century (*BASOR* 215, 33); other sherds from
similar vessels found at Sardis as well as comparative
material[83] seem to be of later date, i.e. fifth to sixth
century. Another, more fragmentary piece, **510**,
from the latrine, is perhaps datable to the sixth or
early seventh century (*BASOR* 157, 34–35). A very
similar bottle section, **511**, was found with Early Byzantine material—fragments of ovoid and cylindrical
bottles, remains of goblets, window panes etc.—in the
drain along the east wall of BE-H (*BASOR* 177, 23)
and is datable to the time after the early fifth century.

Another fragment, **512**, excavated in the area east
of Syn Fc, is certainly fifth or sixth century (*BASOR*
177, 19); much glass came to light close by, including
fragments of a beaker, of goblets, lamps, ring bases
(cf. **401–444**), and pattern-molded ware. The section
of a large bottle, **513**, of the type under discussion
has, unfortunately, no provenance.

Two bottle necks, **515** and **516**, very similar to those
just described, but of relatively thick glass, may have
come from identical or at least very similar bottles.
This seems to be confirmed by the angle of inclination
of the wall of the neck and the estimated diameter of
the rims, about 7 cm. Found in 1959 they come from
BS and are certainly of Early Byzantine date.

509 *Pls. 14, 26*. G72.1:8181. Squat ovoid body with concave base, wide conical neck slightly spreading at rim;
mended, half of neck missing. Amber.
H. 7.7, diam. rim 7.5, body 7.1.
PN/E grave 72.2 (below the floor) under skull *89.50, see
BASOR 211, 18; 215, 33–37. (On p. 33 of *BASOR* 215 the
inventory number is incorrectly listed as G72.2:8181).

510 G59.71:2217. Almost spherical with concave base,
wide conical neck fractured above shoulder constriction.
Fabric 2, eggshell.
P.H. 7.3, max. diam. 6.3.
BE-H latrine W51-W52/N4-N11 *97.00-96.00.
AJA 66, 11, no. 14, pls. 7:17, 10:14.

83. A close parallel: Martini (supra n. 80) no. 1:42, fig. 31, found
with coins of A.D. 600–620 and 650–670. For a neck which probably formed part of a similar bottle and is datable to the 6th C. or
later cf. Harden (supra n. 81) 63, no. 25, fig. 37. For related late
and post-Roman bottles with wide, conical neck cf. Barag, "Glass
Vessels," pls. 5, 7, 23. Cf. also Langdon and Harden (supra n. 50)
fig. 5, nos. 29ff. (5th–6th C.). An early Islamic example with higher
body than the bottle from Sardis, datable perhaps to the 7th or 8th
C. cf. C. J. Lamm, "Les verres trouvés à Suse," *Syria* 12 (1931) pl.
75:1. For another Islamic example, although with added handle,
cf. Saldern, *Slg. Hentrich*, no. 393.

511 Portion of concave base and lower body. Fabric cannot be determined; dirty fire scum.
Est. diam. base 8.
BE-H 1964 E11-E12.1/N42-N44, inside large drain along wall of MC,*96.50-96.00.

512 One half of wide, conical neck curving to shoulder. Fabric 3.
P.H. 3.6, diam. lower neck 2.7, Th. 0.2.
E of Syn Fc 1964 E128-E130/N4-N6 *97.00-96.75.

513 *Pl. 26.* Wide conical neck curving to shoulder. Fabric 1.
Est. diam. lower neck (at constriction) ca. 14.
Chance find.

514 Section of upper body of bottle. Fabric 3.
P.H. 4.2, est. diam. 5.8, Th. 0.15.
E of BS E 19, 1963 E122.5-E123/S8.5-S8.6 *97.30-96.75.

515 Upper portions of conical bottle necks. Fabric 1, thick.
P.H. up to 4.5, est. diam. ca. 7.
BS W 13, 1959 W53-W56/S1.50-S5.25 *98.00-96.75.

516 BS E 5, 1959 E35-E40/S4 *98.50-97.00.

Type 8: Bottles with Constriction below Wide Rim

A single neck fragment has a wide, folded rim with a constriction below; the body may have been spherical or oval-ovoid (**517**).[84] More sherds of this bottle type are no doubt among the finds but could not be identified. The stratification is uncertain; it comes from the late Roman-Early Byzantine bath (*BASOR* 166, 18) at PN but its general appearance—shape, fabric 1 etc.—is more in keeping with Byzantine than late Roman Imperial glass; and a knobbed goblet stem was found nearby at a slightly lower level. It might be datable to the fifth century.

517 *Pl. 27.* Rounded shoulder of spherical(?) bowl (or globular or cylindrical bottle), infolded rim. Fabric 1.
P.H. 5.5, est. diam. rim 8, Th. 0.2.
PN 1961 W250/S370 *98.58.

Type 9: Bottles with Bulging Body

This group is represented here by only one example, **518**; others may not have been recognized. According to the wall section recorded here the vessel had a body with pronounced bulge and a wide neck, the lower part of which tapers upward. Found in the bath at PN with other Early Byzantine material such as a fragmented, pattern-molded bowl lamp (**237**)

and a conical neck of a bottle (**487**), it is probably of fifth century date (*BASOR* 166, 17).

518 *Pl. 27.* G61.23:3775. Wall fragment.
P. dim. 5.5, est. diam. 8, Th. 0.15.
PN W265/S360 *88.45-88.33. Found with **237** and **487**.

Type 10: Bottles with Inner Constriction at Lower Neck

The shape of these bottles is identical to that of pattern-molded bottles, **643–646**: an ovoid body with concave base and neck tapering upward. The most characteristic feature is an inner constriction or diaphragm at the lower neck at the point where the neck joins the shoulder, preventing quick evaporation of the liquid. This diaphragm, which generally has an opening of about 0.3 to 0.7 cm. in diameter, was fashioned before the neck was drawn out and shaped.[85]

As only the bottle sections with the constriction were recorded a safe reconstruction of the entire vessel is most difficult. The most diagnostic section of such a vessel is the thick part of the wall where body, neck, and diaphragm form a unit. As such fragments are easily recognizable, we assume that not many have been left unnoticed. As only relatively few of them were recorded, this type does not appear to have been common at Sardis, perhaps due to the difficulty of manufacture.

The most complete example, **519**, has a neck which measures, up to its fracture, 9.5 cm.; entire necks must have extended to 10 to 15 cm. A complete bottle appears to have measured about 20 to 30 cm., equal in height to the pattern-molded examples (cf. **643–646**).

All five specimens recorded were found in BS E 13 (*BASOR* 170, 50–51). This shop, rich in glass of all kinds, seems to have been in use particularly in the sixth century (*BASOR* 174, 45). One fragment, **521**, was found together with the portion of a similar, pattern-molded bottle, **643**. The last one, **523**, was discovered in 1967 when exploring a drain leading from

---

84. For bottles of this type, already popular in late Roman glass, cf. esp. Barag," Glass Vessels," pl. 6 et passim.

85. Parallel pieces are difficult to identify both in publications without profile drawings and in museums, where, if one cannot handle the objects in question one cannot tell whether or not they have diaphragms. Constrictions at the lower neck with an inner diaphragm were common in Near Eastern pear-shaped bottles particularly of the 3rd–5th C.; cf. Clairmont, nos. 487ff., pl. 12; *Smith Coll.*, nos. 312ff. For late and post-Roman material, esp. bottles with narrow, cylindrical neck and constriction cf. Barag, "Glass Vessels," pls. 41–42.

BS northward under Syn and Pa (*BASOR* 191, 29). At a certain time, perhaps within the sixth century, BS E 13 may have specialized in selling constricted bottles or even specialized in selling a certain liquid which was available in such bottles.

519 *Pls. 14, 27.* G62.11:4487. Shoulder and neck of bottle. Rounded shoulder, neck tapering upward, diaphragm at lower neck int.; mended. Fabric 1.
P.H. 11, of neck 9.5; diam. neck 3.8–4.5, of opening in diaphragm 0.5; Th. 0.2–0.3.
BS E 13 E77/S2.80 *96.60-96.50.

520 One-half of lower neck. Fabric 1.
P.H. 5; diam. lower neck 4, of opening in diaphragm 0.5; Th. 0.2–0.4.
BS E 13, 1962 E75-E80/S0-S3.5 to *96.50 floor.

521 Lower part of neck; fire-softened. Fabric 2.
P.H. 6; diam. 4.2, of opening in diaphragm 0.3; Th. 0.2–0.3.
BS E 13, 1963.

522 Lower neck and portion of shoulder, similar to **519** but with double-walled diaphragm. Fabric 1; dirty fire scum.
P.H. 4.
BS E 13, 1962 E76-E77/S2.80, yellow fill, *96.60-96.50.

523 Lower part of neck. Fabric 2; heavily corroded.
P.H. 4; diam. 3.6, of opening in diaphragm 0.7; Th. 0.2–0.3.
BS E 13, 1967 E75-E78/S0-S2 *95.36.

Type 11: Concave Bases

Among the bases most common in ancient glass are those having a concave underside, i.e. the lowest portion of the vessel wall turns more or less gently inward to a slightly pushed-up base; some show a pronounced kick. This base construction was popular in Roman as well as in Byzantine and Islamic glass. Such a base found without a body section is practically impossible to date by its appearance alone: the base of an ordinary bottle or bowl without ring made in the second century is in most cases indistinguishable from one made in the fifth or tenth century.[86]

The Early Byzantine glass finds from Sardis include virtually hundreds of concave bases or the remains thereof. Undoubtedly they originally formed parts of bottles of various shapes (cf. **476–523**), including those with spiral thread decoration (cf. **607–621**), as well as of beakers (cf. **472–475**), bowls, and perhaps other vessels. As not a single base is preserved that shows a wall section higher than about 3.5

cm. at the fracture, no conclusions can be drawn as to the original vessel types of any of the sherds recorded. They have, therefore, been grouped together.

All have slightly curving walls which may have continued into an ovoid, oval, or cylindrical body. The base diameter ranges from ca. 2.5 to 10 cm. Two-thirds to four-fifths are of fabric 1; the second most frequent fabric is fabric 3.

The base fragments are listed in a statistical fashion to show that they are frequent and of Early Byzantine date. In some cases sherds are recorded because of certain characteristics that seem to be particularly interesting.

A few of the bases show faintly engraved lines at the lower body (**524, 526**). On comparing them to Roman Imperial bottles and beakers of the fourth century with engraved lines[87] one might conclude that these bases stem from bottles, rather than beakers, as the latter do not tend to have the engraved lines. Of four bases with engraved lines, **524** and **525** come from BE-B and BS E 13 respectively, while the others have no known provenance (**526**) or are unstratified (**527**).

Vessel bases that show no wheel-engraved lines come from various areas within the Syn and B complex and from HoB. It is conceivable that some of these bases, especially those discovered in upper or unstratified levels at HoB, might be late Roman. **528** is from BS E 12, **529** from Syn Fc (found with a ring base, cf. **401–444**, and cullet). A base from the top level of the Lydian Trench at HoB (**530**), is perhaps datable because it was found with a definitely Byzantine ring base (cf. **401–444**) in fabric 3 glass and various rim fragments of typically Early Byzantine ware. Two others, **532** (found with a lamp base of type 3) and **533**, come from Pa-W. Discovered with bottle necks with spiral thread (**616**) at BE-H, **534** seems to have been part of a cylindrical bottle. **535** and **536** were found at HoB and they are either late Roman or Early Byzantine.

A series of relatively flat bases that curve gently upward are reminiscent of bulbous bottles with similar bases (cf. **509–516**). They come from the fill under the sidewalk in front of BS E 14 (**537**), BS E 13 (**538**), Pa-S (**539**), the area west of unit L at PN (**540**, either late Roman or Byzantine; *BASOR* 174, 20ff.), east of Syn (**541**), B-W N area (**542**), and Byzantine graves at

---

86. For an approximately contemporary series of concave bases from Jerash cf. Baur-Kraeling, 527ff., fig. 23. For a series of 8th C. bases cf. Negro Ponzi (supra n. 52) 86, nos. 81ff.

87. Isings, nos. 51, 106.

HoB, (**543**). **541** and **543** were found with ring bases (cf. **401–444**, esp. **437**) and lamps of type 3. **544** comes from a tank in a late Roman or—more likely—Early Byzantine context in PN.

Concave Bases

524 One fourth of base curving to cylindrical(?) lower part of vessel; one faint engraved line 2 cm. above base. Fabric 1.
P.H. 2.9, est. diam. 9.
BE-B 1966 E16-E18/N8-N9 *97.00-96.50.

525 One half of base. Yellowish-green.
P.H. 2, est. diam. 5.5.
BS E 13, 1962 E76-E79/S2.5-S3.5 *96.40-96.10.

526 Part of base; one engraved line. Fabric 1.
P.H. 3.8, est. diam. 9.
Chance find 1961.

527 Part of base. Fabric 1.
Est. diam. 5.7.
MTE 1964 *112.50.

528 Lower part of probably cylindrical bottle or breaker(?). Fabric 1(?); dirty fire scum.
P.H. 2.3, diam. 4.8.
BS E 12, 1964 E70-E75/S1-S3 *98.25-98.00.

529 With a portion of slightly conical wall. Dirty fire scum.
Diam. 3.5.
Syn Fc 1967 E99.25-E100/N5.50-N15.20 *96.31.

530 Fabric 2.
HoB 1961 E0/S90-S95 *100.40-100.10.

531 Fabric 1.
Diam. 5.
MTE 1964 N end to *111.00.

532 Fabric. 4.
Diam. 5.
Pa-W 1967 E36.80-E43.80/N95.70-N104 *99.80-96.36.

533 Fabric 1.
Diam. 4.5.
Pa-W 1964 E33.50-E35/N57-N58.50 *92.82-92.62.

534 *Pl. 27.* Lower portion of cylindrical vessel. Fabric 2, eggshell.
P.H. 3.7, diam. 5.
BE-H 1964 E11-E12/N42-N44 *96.50-96.00.

535 Pushed-up base. Fabric 3.
Diam. ca. 5.5.
HoB 1962 W10/S85, near yellow earth, to *99.80.

536 Fabric 4.
Diam. ca. 6.5.
HoB 1963 W20/S110-S115 to *101.00.

537 Fabric 1.
Diam. 6.
RTE 1962, fill under sidewalk in front of BS E 13, *93.70.

538 Fabric 2, very thick.
Diam. ca. 10, Th. to 0.9.
BS E 13, 1962 E75-E76/S2-S5 *96.70-96.20.

539 Fabric 1.
Diam. 6.
Pa-S 1965 E80-E82/N20.8-N23.9 *95.45.

540 *Pl. 27.* Dirty fire scum.
Est. diam. 9.
PN 1963 W250/S340 *87.50-87.00.

541 Fabric 2; black scum, fire-twisted.
Diam. ca. 10.
E of Syn 1963 E117-E118.9/SI-N2 *97.50-96.75.

542 Fabric 1.
Diam. ca. 7.
B-W N Area 1966 W29-W30.5/N89.7-N90.3 *95.80-95.40.

543 Fabric 1.
Diam. 8.
HoB 1962 E5/S110-S115, Byzantine graves, *102.00.

544 *Pl. 27.* Fabric 4; dark scum.
Diam. ca. 5.5.
PN 1963, W230-W234/S347-S350, tank b 1–2 *87.65-87.25

545 Fabric 1.
Diam. 8.
HoB unit 18 1959 E30/S60 *96.50-96.25.

Type 12: Thick-Walled Fragments

Plain and pattern-molded bottles were made in thin or medium thick as well as in very thick glass, the latter generally fabric 1. A few sections from the body of thick-walled vessels should receive particular attention as this heavy glass seems to have been the exception in Sardis. The shape of the original vessels is unknown as the sherds are too small for reconstruction. One fragment comes from east of Syn Fc (**546**), two others were found at Pa (**547** and **548**); all are no doubt of Early Byzantine date.

546 E of Syn 1963 E117-E119/N5-N6.70 *96.75.

547 Pa-S 1966 E40-E43.50/N27-N30 *97.00-96.00.

548 Pa-S 1965 E92-E93/N25 *97.00.

Type 13: Tube-Shaped Miniature Bottles

Perhaps the most inconspicuous vessels found in Sardis are small, tube-shaped vials and miniature ampullae of varying size (cf. **560–571**). Tube-shaped bottles of similar form are rare in late Roman glass.[88] Vessels almost identical in shape to those found in Sardis continued to be made after the seventh century; they were fairly frequent in late and post-medieval periods in the West.

The Sardis excavations yielded no complete specimen; in each case the upper section is missing. By analogy with other vessels, however, the cylindrical bottles almost certainly had a neck of slightly concave profile flaring to a plain lip. The lower portion of such a bottle (**549**) was among the finds in BS E 13 (*BASOR* 170, 50). A second piece (**550**) comes from the Early Byzantine level of BE-H (*BASOR* 166, 46ff.). The lower portion of a third bottle (**551**) from MTE was unstratified but was found with glass mostly datable to the Early Byzantine period, including goblet feet, ring bases (cf. **401–444**), remains of lamps etc. A fourth example (**552**), found at PN, may be Early or Middle Byzantine (*BASOR* 177, 3). In addition, another fragment (**553**) was found in grave 67.3 at PN and may be either very late Roman Imperial or Early Byzantine (*BASOR* 191, 11).

A series of lower bodies of probably tubular bottles have concave or pushed-up bases; they either belonged to bottles of the type just discussed, or to very similar vessels. While two (**554** and **555**) are unstratified, three others are securely dated within the Early Byzantine period. **556** comes from Syn (*BASOR* 174, 30ff.); with it was found part of the bottom of a turquoise blue glass bowl (**392**). One of the very few pieces made of blue glass (fabric 6) is **557**, found in BS E 13. **558** was excavated in BS E 15 with the base of a bowl of greenish glass (*BASOR* 170, 51). The upper neck and out-turned, folded rim of a small bottle (**559**) may have been part of a cylindrical bottle. It comes from W of B and should be associated with the Early Byzantine level although a later, Middle Byzantine date is conceivable. Among the fragments found with it are

sections of a salver (**381**), a cup base (cf. **401–444**), and a bottle (**496**; *BASOR* 191, 38).

Bases of Cylindrical Bottles

549 *Pl. 27.* Cylindrical, pushed-up base. Fabric 1.
P.H. 1.8, diam. 2.7.
BS E 13, 1962 E75-E80/S0-S3.5 to *96.50 floor.

550 Fabric 2(?); heavy scum.
P.H. 2.2, diam. 1.5.
BE-H 1961 E15-E20/N70-N80 *97.50-96.00.

551 Fabric 2(?); heavy scum, dirt.
P.H. 1.4, diam. 2.
MTE 1964 E60-E65/S149-S153 *110.00-109.50.

552 *Pl. 27.* Cylindrical with pushed-up base; mended. Fabric 2(?); corroded, heavy scum.
P.H. 3, diam. 2.5.
PN 1964 W254-W255/S339-S340 *88.20-88.10.

553 *Pl. 27.* Fabric 2; thick, devitrified, white scum.
P.H. 2.2, Th. 0.5.
PN grave 67.3, 1967.

Concave Bases

554 *Pl. 27.* Green, thin; iridescent, brown scum.
P.H. 1, diam. 3.8.
Chance find 1961.

555 Fabric 4; slight weathering.
Diam. 3.5.
AcT 1962 W0/N22 *404.20-403.90.

556 Eggshell; scum, corroded.
Diam. 2.3.
Syn MH 1963 E55-E56/N3-N3.5 *96.50-96.00. Found with **392**.

557 Fabric 6 (blue), eggshell.
Est. diam. 2.3.
BS E 13, 1962 E75-E76/S2-S9 *96.70-96.20.

558 Fabric 1; thick.
Diam. 2.5.
BS E 15, 1962 E87-E90/S1.5-S3.5 *95.00-94.00, fill under floor.

---

88. Cf., for example, the material assembled by Barag, "Glass Vessels," passim; Harden, *Karanis*, no. 696, pl. 19. Tube-shaped bottles of Sassanian date, i.e. ca. 4th–6th C., were found in Ctesiphon (preserved in The Metropolitan Museum of Art, New York, Dept. of Ancient Near Eastern Art).

559 *Pl. 27.* Three fourths of cylindrical neck with folded out-turned rim. Fabric 1.
P.H. 1.5, diam. 3.5.
W of B 1967 trench C W100-W102.70/N12.20-N14.20 *96.80-96.25. Found with **381** and **496**.

Type 14: Conical and Spherical Miniature Bottles

The bottle form of this group is a short, conical body having a concave base, a long, tube-like neck, and, generally, a folded rim. The prototype can be found in numerous bottles of Roman date.[89] Among the best preserved examples at Sardis is a bottle from BS W 2 (**560**) which does not appear to be well stratified but is certainly either very late Roman or Early Byzantine. The lower portion of an identical piece (**561**) is unquestionably Early Byzantine as it comes from BS E 13 (*BASOR* 170, 50). A third example, **562**, with a more bulbous body, comes from BE-A (*BASOR* 187, 10–12, designated Unit A').

The following examples have smaller, more rudimentary bodies and even longer necks. From BS W 1, **563** is datable to the fifth or sixth century (*BASOR* 154, 16–18). A well-preserved example, **565**, having a bulbous body and a flaring upper neck, comes from the upper level at BE-CC. However, it has all the appearance of Early, rather than Middle, Byzantine glass (*BASOR* 187, 56–57). Among the other glass finds in this area were fragments of salvers, bowls, cups, and beakers, all datable to a period after the early fifth century.

A series of bottle necks that are likely to come from identical or very similar bottles are so fragmentary that it is difficult to assign them, by their shape alone, to a specific period. A bottle neck (**566**) from UT could perhaps be associated with fifth century material according to the excavator, G. Métraux, although it is not precisely stratified. However, among the glass finds nearby were sections of vessels with facet-cut decoration and with warts (cf. **62–76** and **81–90**) of late Roman date (*BASOR* 157, 19ff.). **567** was found in LNH 1 with fragmentary beaker lamps and is certainly Byzantine.

Closely related miniature bottles include a short bottle of concave, tube-like shape (**568**) from BS W 8, certainly Early Byzantine (*BASOR* 157, 32–33) and a tiny, pear-shaped bottle (**569**) from the neighboring BS W 9. A small vial with spherical body (**570**) may belong to the time just before the date assigned to the *96.60 mosaic floor in BE-B which appears to have been a work of the fourth or fifth century (*BASOR* 187, 16). A bottle neck with spiral thread (cf. **607–621**) was found close by at about the same level and

one should link them with the other numerous Early Byzantine finds from this room.

Most interesting is a small vial with flattened, oval body in a tube-shaped bronze casing or wrapper (**571**). It has been suggested that vessels enclosed in such a metal tube might have been "curse" bottles. Found in Pa-S at the stylobate level, it certainly belongs to the fifth or sixth century (*BASOR* 191, 28–29).

560 *Pl. 15.* G58.15:466. Sloping body with pushed-up base, cylindrical neck, infolded rim. Fabric 2.
H. 5.7, diam. base 3.
BS W 2, ca. 2.20–2.30 m. below top of wall, level III.
Hanfmann, *JGS* 53, n.10.

561 Lower body; pushed-up base curving to pear-shaped or conical body. Fabric 2; eggshell.
Diam. base 3.5.
BS E 13, 1962 E75-E76/S2-S9 *96.70-96.20.

562 *Pl. 15.* G66.1:6975. Bulbous body with pushed-up base, cylindrical neck with infolded rim.
H. 6.6, diam. 2.4.
BE-A E12.32/N0.90 *96.80.

563 G58.60:796. Conical with slightly pushed-up base, cylindrical neck with lower constriction, infolded rim. Fabric 2.
H. 6.5, diam. base 1.5.
BS W 1, over level II, floor 2, fill.
Hanfmann, *JGS*, 53, fig. 3.

564 *Pls. 15, 27.* G62.2:4160. Irregular, sloping with pushed-up base, cylindrical neck, wide mouth with infolded rim. Fabric 5; bubbly.
H. 7, diam. 1.7.
RTE E20-E22/S6-S8, in fill over sidewalk, *97.25-96.75.

565 *Pl. 15.* G66.10:7261. Slightly flattened, bulbous body with pushed-up base, flaring neck with infolded rim. Fabric 2; black scum.
H. 5.7, diam. 2.3.
BE-CC E32.92/N99.90 *98.

566 Neck with infolded rim. Fabric 5.
P.H. 5, diam. rim 1.3.
UT 1959 E90/S210 *123.45-122.75.

89. Isings, nos. 82, 83. Closely related in shape and time to the Sardis material are the 4th–5th C. finds from Jerash: Baur-Kraeling, 537ff., figs. 25ff. Cf. also the late Roman bottles in Harden, *Karanis,* nos 819ff., pl. 20, and the Sassanian bottles in Puttrich-Reignard, 17ff. (cf. M. Negro Ponzi, "The Excavations at Choche [the Presumed Ctesiphon]," *Mesopotamia* 1 [1966] 87, fig. 34). Small spherical bottles with conical necks, datable to the 5th–6th C., from Apamea: Canivet, 64ff., fig. 14.

567 *Pl. 27.* Cylindrical neck with infolded rim spreading at bottom to body. Fabric 5.
P.H. 5.7, diam. rim 1.7.
LNH 1 1968 E40.40-E44.83/N103.26-N110.25, SW corner trench, *96.80-96.00.

568 G59.28:1374. Portions of 2 bottles; fragmentary, tube-like, waisted at center. Fabric 2.
H. 6.5, diam. 2.
BS W 8 W22-W24/S5.25-S6.25 *97.80-97.00.

569 *Pl. 15.* G59.25:1341a. Miniature; oval, flattened, thickened rim. Fabric 2.
H. 4.7, diam. 1.2.
BS W9 W27-W29/S3 *97.75-96.75.

570 *Pl. 15.* G66.3:7012. Spherical with pushed-up base, tapering neck, rim fractured. Clear with yellow green tint; eggshell.
P.H. 3.4, diam. 2.5.
BE-B E17.62/N7.56 *96.52.

571 *Pl. 15.* G67.6:7444. Bottle found in tube-shaped bronze wrapper; irregular, slightly flattened, oval body tapering to neck with infolded rim. Fabric 2; bubbly.
H. 4.2, diam. 1.4.
Pa-S E38.21/N28.77 *96.41.

RIMS OF VESSELS

A large percentage of the rims excavated within an Early Byzantine context are plain, generally having a slightly thickened (fire polished) edge. The degree of inclination and the estimated diameter of the rim often serves as the only means for a hypothetical reconstruction. For example, a plain rim of fabric 1 glass with a fairly low degree of inclination and an estimated diameter of over 10 cm. could have belonged to a salver (**377–381**). An identically formed rim section of which the degree of inclination is fairly steep and the estimated diameter ranges around 6 to 8 cm. may have been part of a goblet. On the other hand, similar rims with larger diameters (8–10 cm.) could also have belonged to bowls (cf. **382** and **383**), or other vessels.

In this catalogue those rim sections that could be assigned to a type are included under that category. Those from vessels of unknown form which warrant special attention are grouped together here. However, at least some of them probably formed part of vessels already referred to in previous chapters.

The rim sections represent only a sampling of the total number of vessel rims found which, although left uncatalogued, have been incorporated in the statistics. They also may serve as indication that the variety of forms of the glass discovered in Sardis may have been considerably greater than a cursory survey of the preserved and reconstructed vessels illustrated in the plates would suggest.

Plain Vessel Rims

Most frequent are plain rims of slightly conical vessels or vessel necks. Rim sections of small diameters could be from conical bottle necks (cf. **476–487**). Those with estimated diameters of ca. 6 to 8 cm. may have been part of goblets (cf. **300–373**) while diameters of over ca. 8 cm. seem to indicate that the rims must have been part of bottles with wide necks or of bowls (cf. **509–516** and **382** and **383**).

Of the rim sections with an estimated diameter of about 8 cm. (**572** and **573**), **572** seems to be a late intrusion in a Hellenistic level (*BASOR* 170, 10). **573** comes from BE-B and, although apparently from below *96.60, should be datable to the fifth century (*BASOR* 187, 16, designated Unit B′). The rim section of a vessel with steep walls, **574**, perhaps a large bottle, is unquestionably Early Byzantine as it was found in the drain of RTE together with the remains of quantities of goblets, plain and threaded bottles, and windows (*BASOR* 170, 38). Another fragment with the same profile but of thicker glass, **575**, is unstratified but was found with portions of a lamp of type 3 and two slightly concave bases of Early Byzantine bottles (*BASOR* 177, 14–15).

A number of other rim fragments (**576** and **577**) were found in BS E 13 (*BASOR* 170, 50) and other areas around Syn. A fragment, **578**, with an estimated diameter of 14 cm. was probably part of a bowl; the excavator dates it to the sixth or early seventh century.

The rim of a rather shallow vessel, **579**, is unstratified (*BASOR* 182, 15ff.), but the color of the glass (fabric 3) and the fact that it was found with the base ring of a large dish (**470**) leaves no doubt that it is post-Roman. The degree of inclination suggests it might have been part of a salver, but as its estimated diameter is only 7 cm. and its thickness considerable, the rim may have formed part of a small, shallow bowl.

582 is from a vessel with slightly convex wall which turns upward at the lip, a bowl or a goblet. It is certainly Byzantine; goblet feet and window fragments were found with it (*BASOR* 170, 49). The fragment, **581**, was part of a vessel which flared slightly at the rim—probably a goblet or a bowl. Its Early Byzantine date is certain as it comes from BS W 4 (*BASOR* 154,

16–17). **582** has a fairly wide opening above a waisted part that seems to represent a wide neck, and comes from an unstratified level at HoB. However, as the base of a thick cup, **186**, and the curved section of a bowl, **201**, of typically Roman form were found with it, its Byzantine date is open.

Another vessel type appears to be represented by a rim section with an estimated diameter of only 3 cm. (**583**). It is conceivable that it was a bottle rim with slightly convex lip. The fragment can probably be associated with the fifth to sixth century level at Syn (*BASOR* 182, 40).

Among the other fragments with slightly convex wall and an edge turning upward or inward are two sherds, one from BS E 17, **584**, and the other from BE-S, **585** (cf. **592**). Close to the first fragment were found other rim sections, including folded rims and remains of pattern-molded ribbed ware (*BASOR* 174, 45). The second sherd belongs to the many Early Byzantine glass finds from BE-S where remains of goblets, threaded necks, lamps, pattern-molded glass etc., were discovered (*BASOR* 182, 30–31); evidence from coins found nearby at the same level points to a date of about the late fifth to early sixth century. The rim fragment of a small bowl, or lamp (**586** with slightly flaring upper wall and a lip sharply turned upward comes from BE-B, a room rich in Early Byzantine glass (*BASOR* 187, 15).

Three plain rim fragments of blue glass are listed here (**587–589**). Their size is so small that even an attempt to guess the shape of the original vessel would be highly speculative. Nonetheless, they could be parts of goblets, a category generally carefully executed in thin, well-made ware. Two of the three fragments can be dated safely within the Early Byzantine period. **587** comes from the north-south street between BS W 13 and W 14, underneath the destruction level of A.D. 616 (*BASOR* 186, 28ff.). **588** was found at Syn Fc and is datable to the fifth to sixth century (*BASOR* 174, 43–44); remains of globlets, lamps, and bottles as well as glass cakes (cf. **729–737**) were found at the same location.

Two rim sections, perhaps of bowls (**590** and **591**), have a waisted profile, probably representing a neck constriction below a wide lip. The first was found in BE-B, just beneath the *96.60 level, associated with ca. A.D. 400 (*BASOR* 187, 16); it is probably datable to the very late fourth or early fifth century. With it came the section of a handled bowl lamp, perhaps of type 1, and two ring bases (cf. **401–444**), an indication that the group is Early Byzantine. An identical rim section in fabric 1, having an estimated diameter of 8 cm., was found in 1962 but has no provenance.

A third sherd, **591**, of very similar shape and fabric, comes from an unstratified layer at MTE with much predominantly Byzantine material, including the remains of goblets, ring bases (cf. **401–444**), lamps of type 3 etc., but also some Roman glass. The rim seems to be post-Roman, however.

572   Fabric 1.
Est. diam. 8.0, Th. 1.5.
HoB 1962 W10/S85 *99.80.

573   One half of rim, crudely made, large bubbles. Fabric 1.
P.H. 5.6, est. diam. 7, Th. 0.1.
BE-B 1966 E16.5-E18.5/N4-N8.7 *96.30-96.10.

574   Almost vertical wall. Fabric 2; eggshell.
P.H. 2.5, est. diam. 8, max. dim. 7.
RTE 1962 drain *95.00.

575   *Pl. 27.* One half of almost vertical rim wall. Fabric 1; thick.
P.H. 5.2, est. diam. 4.5, Th. 0.3.
HoB 1964 E64-E71/S155-S160 *109.80-109.10.

576   *Pl. 27.* Two rims, Fabric 1.
Est. diams. 5.5 and 8, Th. 0.2.
BS E 13, 1962 E76-E79/S2.5-S3.5 *96.40-96.10.

577   *Pl. 27.* Fabric 1; badly made.
Est. diam. 5, Th. 0.25.
BS E 13, 1962 E76-E79/S2.5-S3.5 *96.40-96.10.

578   G59.21a:1328. Fabric 1.
Est. diam. ca. 14.
HoB area 11.
AJA 66, 10, no. 11a, pl. 9:11a.

579   *Pl. 27.* Rim section of shallow vessel. Fabric 3; thick.
P.H. 5.7, est. diam. 7, Th. 0.15-0.2.
HoB 1965 upper mixed fill, *102.40-100.50. Found with **470**.

580   *Pl. 27.* Rim section of slightly convex vessel. Fabric 1; eggshell.
P.H. 5.2, est. diam. 9.
RTW 1962 E30-E35/S34 *96.00.

581   G59.16:1262. Slightly flaring rim section. Fabric 1; eggshell.
Est. diam. ca. 9.
BS W 4 *98.00-97.50.
AJA 66, 11, no. 11b, pl. 9:11b.

582   One half of flaring rim tapering to waisted neck. Fabric 3.
P.H. 3.7, est. diam. 8, diam. at neck 4.

HoB 1962 E5-E10/S115 to *101.40. Found with **186** and **201**.

583 Slightly convex rim section Fabric 2.
Est. diam. 3, Th. 0.2.
Syn 1965 E90-E93/N1.5-N3 *95.60-95.50.

584 *Pl. 27.* One third of rim of slightly convex upper part of vessel. Fabric 1; eggshell, iridescent, white scaling scum.
P.H. 6.5, est. diam. 9.
BS E 17, 1963 subfloor *96.40-95.80.

585 Rim section of probably conical vessel with edge turned inward. Fabric 1; heavy scum.
Est. diam. 8, Th. 1.5.
BE-S 1965 ca. E25.75/N27.50 *96.10.

586 *Pl. 27.* One fourth of slightly flaring rim, edge turned upward and thickened. Fabric 1, eggshell.
P.H. 4.7, est. diam. 7.
BE-B 1966 E16.5-E18.5/N4-N8.7 *96.30-96.10.

587 Small rim fragment, slightly convex. Fabric 6 (blue).
P.H. 1.2, Th. 0.1.
Unit W of BS W 13, 1966 W59-W63/S3-S6 down from *97.00-96.50.

588 Small fragment with curvature, perhaps from bottom of bottle. Fabric 6? (deep blue).
P.H. 1.5, Th. 0.1-0.2 (at bottom).
Syn Fc 1963 E110-E115/N1.5-N6 *97.50-97.25.

589 Convex. Fabric 6 (blue).
Est. diam. 9.
Chance find.

590 *Pl. 27.* One third of short, flaring rim above constricted neck. Probably fabric 1.
Max. dim. 7.2, est. diam. 9, Th. 1.
BE-B 1966 E16.50-E18.50/N4-N8.70 *96.40-96.10.

591 *Pl. 27.* Short, curved rim above constricted neck. Fabric 1; corroded, iridescent, brown scum.
Max. dim. 5, est. diam. 7, Th. 1.
MTE 1964 E60-E65/S149-S153 *110.00-109.50.

Folded Rims of Vessels

Vessels with rims folded inward appear to have been less frequent than those with plain rims. The rim fragment of a bowl or cup with conical wall and a folded rim turned inward (**592**, cf. **584** and **585**) comes from BE-C and may be datable to a period after the early fifth century (*BASOR* 187, 19–20; designated unit C'). In the same room were found other rim fragments and vessel sections probably of bowl lamps of type 2. The most numerous series of folded rims comes from vessels with walls flaring or conical at top, similar to those mentioned above (cf. **581**). Of two recorded here, **593** comes from near BS W 2 and is probably of sixth century date; this shop contained much glass, including the remains of goblets and windows (*BASOR* 154, 18). The other sherd (**594**) was found in BS E 5 (*BASOR* 191, 17–18) together with several fragments of glass vessels, including part of a vessel wall with an applied, straight thread. The material found there is probably contemporary with that of the neighboring shops and thus datable to the fifth or sixth century.

Vessel rims with estimated diameters of ca. 6 or 7 cm. seem to have been part of small bowls or bottles with conical and funnel necks. Several are recorded (**595–597**) from BS E 13 (*BASOR* 170, 50). Coins found at the same level of RT as **598** suggest it is probably of a date in the early seventh century (*BASOR* 166, 43–44).

Of rims with slightly flaring walls and estimated diameters of 14 to 18 cm., **602** comes from BS E 17 and was found with a handle of a bowl lamp (**264**) and the remains of goblets, pattern-molded vessels etc. **600** and **601** are unstratified as they come from AcT where they may be associated with the first Byzantine occupation in the sixth century (*BASOR* 170, 31–32).

The rim of a bowl with vertical, straight walls at the upper section (**603**) was excavated in BS E 16, an area rich in Byzantine glass, namely the remains of lamps, salvers, bottles, and cups as well as cullet and tesserae (*BASOR* 174, 45). Probably from a bottle or vase, **604** is from the upper level of the Lydian Trench at HoB and seems to be Early Byzantine rather than Roman Imperial, judging by its color (fabric 1) and its—admittedly hypothetical—shape, namely a globular vessel with wide neck (*BASOR* 166, 5ff.). A deformed conical neck (**605**) comes from Syn Porch where numerous fragments of Early Byzantine glass came to light: parts of beakers, lamps, goblets, cups etc.

Finally, the folded rim of a shallow vessel, probably a plate or salver (**606**) with an estimated diameter of 11 cm., was found at LNH; it may be contemporary with the destruction of the Middle Byzantine furnace. This rim is curious in that it is pinched at regular intervals to form air traps. It is possible that this rim is of a period succeeding the Early Byzantine occupation level, Islamic or Middle Byzantine.

592 Conical wall with folded rim turned inward. Decolorized, black scum.
P.H. 5, est. diam. 7, Th. 0.1-0.2.
BE-C 1966 E18-E19/N11.7-N12.5 to *96.15.

593 G58.95:1048. Portion of flaring wall and folded rim with wall fractured at lower end where it turns toward the vertical. Fabric 2, thick.
Est. diam. 11.5, Th. to 0.45.
S of BS W 2 *96.50.
AJA 66, 11, no. 17, pl. 10:17.

594 *Pl. 27.* One sixth of vessel (bowl?) with slightly flaring wall, folded rim. Probably fabric 1.
P.H. 4.5, est. diam. 9, Th. 0.1.
BS E 5, 1967 E34-E36/S3-S5 *97.20-97.00.

595 *Pl. 27.* Conical wall with one third of heavy, folded rim. Fabric 1, thick.
P.H. 4, est. diam. 7, Th. to 0.3.
BS E 13, 1962 E75-E80/S0-S3.5 to *96.50 floor.

596 Same as **595**, three fourths preserved. Dirty fire scum.
Est. diam. 6, Th. 0.35.
BS E 13, 1962 E76-E79/S2.5-S3.5 *96.40-96.10.

597 One-half section of rim similar to **595**. Fabric 1.
Est. diam. 7.5, Th. 0.35.
BS E 13, 1962 E75-E76/S2-S9 *96.70-96.20.

598 Small section of folded rim with conical or flaring wall. Fabric 3.
Max. dim. 3.5, est. diam. 12.
RT 1961 E2-E9/S28-S31 *97.30-96.50.

599 Similar to **594**. Fabric 1.
Est. diam. 14.
AcT 1962 E6-E10/N19-N20, level Ib, unit 6, near E-W wall, *403.00-402.40.

600 Similar to **599**. Eggshell.
Est. diam. 14.
AcT 1962 E6-E9/N17-N18 *402.50-401.00.

601 Identical portions of rims in fabric 2.
Est. diam. 18.
Same provenance as **600**.

602 Fabric 1.
Est. diam. 18.
BS E 17, 1963 E97-E101/S1-S4, clearing window *96.60-95.90.

603 *Pl. 27.* One fifth of folded rim with vertical wall. Probably a large, deep bowl. Fabric 2.
Max. dim. 6.5, est. diam. 20, Th. to 0.1.
BS E 16, 1963 E92-E96/S0-S2 *97.50-97.25.

604 *Pl. 27.* One fourth of conical neck flaring at bottom to shoulder(?) of vessel. Neck double-walled because upper portion was infolded, air trap at rim. Fabric 1, thick.
P.H. 6.3, est. diam. 12, Th. 0.2-0.4.
HoB 1961 E5/S95-S100 to *100.40.

605 Conical (deformed) neck with folded rim. Fabric 2; fire-twisted.
P.H. 5, est. diam. 8-10, Th. to 0.25.
Syn Porch 1963 E117-E119/N5-N6.70 *96.75.

606 *Pl. 27.* One fifth of rim, fractured at wall, pinched at regular intervals to form air traps. Probably yellowish; brilliant iridescence.
P.L. 5, est. diam. 11, W. 0.5, Th. to 0.1.
LNH 1, 1968 E44.83-E46/N116.50-N121.01 *97.00-96.30.

VESSELS WITH THREAD DECORATION

Threaded Necks

Of the numerous fragments decorated with applied threads, the most common type is a bottle with applied spiral thread wound around its conical neck. Some of the neck sections are complete down to the point where the neck curves to the shoulder; however, none of the body sections of such a bottle could be identified.

The neck generally has a plain rim and seems to have measured about 6 to 9 cm. to the shoulder; the diameter of the rim varies from 2.2 to 3.0 cm. A thin spiral thread is applied to the upper neck. Judging by parallel examples found at other sites, the body of this bottle type was oval or ovoid, i.e. wider at the shoulder, with a gentle curve toward the base which certainly was concave, very similar to plain bottles (**476-487**). A large number of concave bases of which some are catalogued (cf. **524-545**) may have belonged to bottles with threaded necks. Accordingly, the bottles must have reached a height of ca. 12 to 18 cm. The fabric, often difficult to identify because of heavy weathering or exposure to fire, appears to have been predominantly greenish (fabric 2).

The bottle derives from late Roman Imperial bottles which often have either conical necks or cylindrical necks with a wider, funnel-shaped orifice. Close parallels to the Sardis bottles are afforded by vessels found at Khirbet el-Kerak, datable to the early fifth to early seventh century,[90] Jerash,[91] and other late and post-Roman sites.[92]

---

90. Delougaz-Haines, pl. 59:1, 3.
91. Baur-Kraeling, 533ff., fig. 31 (decorated mostly with threads of blue glass).
92. Cf. Barag, "Glass Vessels," pls. 5, 11 (here *Pl. 25*), 14, 27, 29, 42ff (Pl. 43, here *Pl. 25*). Other sites where bottles of this type were found include Beth Shan (ca. 4th C.; FitzGerald, 42, pl. 39); Jeleme (2nd half 4th C.; Perrot [supra n. 77] 258.3, fig. 3a; Goldstein, pl.

Almost all necks recorded come from safely dated levels of the Early Byzantine period; they are usually intermixed with other, typically Early Byzantine glass datable to the early fifth to early seventh century. Within this period the basic form does not seem to have been subject to change. Two examples (**607** and **608**) belong to levels of the early fifth to early seventh century. A few others (**609–611**) were found above the mosaic floor at Pa-W and appear to be of the fifth or sixth century (*BASOR* 191, 33); other finds in this area include goblet stems, a salver foot, the base of a cylindrical (?) bottle similar to **528** as well as plain bottle necks.

While a find from HoB, **612**, is not stratified (*BASOR* 182, 15–16), another neck, **613**, was found among the fallen columns between BE-H and MC (*BASOR* 177, 21–23). As it was found above the opus sectile floor in MC it may be dated after the fifth century. A third find, **614**, from W of B certainly belongs to the period before the destruction in A.D. 616.

Other bottle necks include three from BE-S (**615**; *BASOR* 182, 30–31) and BE-H (**616**), datable to the time after the renovation in the late fifth century (*BASOR* 177, 21–23); with the latter came portions of pattern-molded ware. A multitude of vessel fragments belong to the same period, including many necks (**617**); they were found in the debris of Pa-W (*BASOR* 162, 40–42) among which are included the remains of goblets, plain vessel rims, and concave bases of bottles; some of the latter perhaps were parts of bottles with threaded necks. Similar material, including a neck (**618**), was also found at MRd (*BASOR* 166, 40). **619** comes from BE-B where quantities of Early Byzantine glass were found; although part of a slightly lower level (*96.40), it certainly must belong to the other, post-Roman glass. Finally a neck from PN (**620**) no doubt belongs to the period of the Early Byzantine burials (*BASOR* 191, 11).

607   G58.83:1009.
HoB surface.
Hanfmann, *JGS*, 53.

608   G58.68:865.
BS W 3 level II, fill below floor 1.
Hanfmann, *JGS*, 53.

609   Neck with small portion of shoulder. Green tint (fabric 2?), eggshell.
H. of neck 6.
Pa-W 1967 E36.80-E43.80/N95.70-N101 *99.86-98.36.

610   H. ca. 8.
Pa-W 1967 E35.85/N95.70 *97.65.

611   H. 8.5.
Pa-W 1967 E36.80-E43.80/N95.70-N104 *99.86-98.36.

612   *Pls. 15, 27.* Neck with portion of shoulder; irregular lip. Greenish (fabric 2?), eggshell; dark scaling scum.
P.H. 9.1, H. neck 8.
HoB 1965 W25-W35/S120-S130 *101.90-101.60.

613   H. of neck 6.
BE-H 1964 E10-E14/N68-N71 *97.00-96.11.

614   H. of neck 7.
W of B 1967 W107/N23.50 *96.76.

615   BE-S 1965 E18-E20/N35-N37 *96.30.

616   Two necks.
H. 6 and 7.
BE-H 1964 E11-E12/N42-N44 *96.50-96.00.

617   Many fragments of necks.
Pa-W 1960 E33-E35/N38-N42 *97.00-96.25.

618   MRd 1961 E0-E3/S8-S10, fill below marble paving, *96.50.

619   BE-B 1966 E22.30/N9-N9.5 *96.40.

620   PN 1967 W265-W271/S330-S340 *88.30-88.10.

621   Group of conical and almost cylindrical bottle necks (diam. 3–6). Mostly clear; dark scum. Fragments from a globular bottle with conical neck; white spiral thread decoration (P.H. 7, diam. 2.3). Found with **508**, the rim of a lamp of type 1 and goblet feet.
BE-C 1973 E15-E17/N22-N23 *99.0-98.19.

Wide-Necked and Related Vessels

A bottle type identical or very similar to bulbous, wide-necked vessels (cf. **509–516**) but having added thread decoration is represented by a few fragments. The diameter of the upper neck varies between 3 and 6 cm.; in comparison **509** has a rim with a diameter of 7.5 cm. The neck sections recorded here could, however, also have formed part of lamps with three handles of a type found at Saqqara which likewise

20); Shavei Zion (5th–6th C.; Barag, *Shavei Zion,* 65ff., nos. 4ff., fig. 16, here *Pl. 25*); Kharjih (6th C.; cf. Harden, *ArchJ,* fig. 1:I, J). For an example of a spherical, pattern-molded, ribbed bottle of the 5th or 6th C., having a cylindrical neck with a narrow lower and wider upper portion with spiral thread cf. Victoria and Albert Museum, London, inv. no. C 12-1939. For Islamic examples of similar shape cf. Hama (probably 9th C.; Riis [supra n. 52] 60, figs. 173–174.

carries thread decoration.[93] The identification of **622** as a lamp is supported by the fact that it was found with a handle (**263**) of a type characteristic for bowl lamps. They were found in BS E 16 and are unquestionably of the fifth to sixth century (*BASOR* 174, 45); among the glass from this shop are the remains of lamps, bottles, vials, salvers as well as tesserae. The exact findspot of an almost identical neck section (**623**) is unknown but Early Byzantine glass was found inside the large drain along the east wall of BE-H at N42, where the piece appears to have been found. Moreover, a section of a vase-like bottle, probably of very similar shape to the original vessel of which **623** was part, appears to have been discovered in its immediate neighborhood. Various broken goblets, cylindrical bottles, ampullae, concave vessel bases, and windows were part of the finds from this location, all probably dating from the fifth century (*BASOR* 177, 23ff.).

Neck section **624** differs from the foregoing examples in that the thread decoration was more carefully applied and the body wall appears to have spread at a wider angle. It comes from an area at PN that contained Lydian structures (*BASOR* 177, 4, 6). Although not well stratified, its shape and fabric determine that it is certainly Early Byzantine. Sections of two other necks (**625**) are safely datable to the fifth or sixth century as they were excavated in BE-B where Early Byzantine glass was plentiful (*BASOR* 187, 15–16).

The following fragments are too small to contribute to the reconstruction of the original vessels. The first (**626**) has no provenance but appears to be Early Byzantine. The second (**627**) comes from the upper fill of Syn MH and probably belongs to the Early Byzantine period (*BASOR* 174, 30ff.). This piece only shows the sections of two threads; they either were part of a more widely spaced spiral thread, or they represent two (or perhaps more) single threads wound around the vessel's body. **628** has a single thread; it comes from BS E 1 where coins ranging from Arcadius and Honorius (A.D. 395–408) to Phocas (602–610) and remains of salvers, lamps, bottles, and window panes were found (*BASOR* 191, 17).

Another small vessel fragment, **629**, with two threads laid crosswise one on top of the other, is from BS E 13 (*BASOR* 170, 50). Two applied threads running parallel to each other appear on a tiny fragment, **630**, that comes from the area east of Syn Fc and is datable to the fifth or sixth century (*BASOR* 174, 47).

A small fragment found in 1959, **631** has an applied spiral thread of red glass and it appears to have been part of an Early Byzantine (or Islamic?) vessel.

Finally, a vessel fragment with a portion of a thread that may have been applied in a zigzag fashion (**632**) was found in BS E 12, a room rich in glass finds of Early Byzantine date (*BASOR* 177, 19–20).

622   *Pl. 27.* Portion of relatively wide, conical neck curving to shoulder; narrow spiral thread. Fabric 1.
P.H. 3.5, diam. at upper fracture 4, Th. 0.15.
BS E 16, 1963 E92-E96/S0-S2 *97.25-96.50. Found with **263**.

623   Similar to **622**.
P.H. 2.4, diam. at upper fracture 3.2.
BE-H 1964 niche on E wall E12.60/N42.5 *97.60, floor level.

624   *Pl. 27.* G64.9:6558. Conical mouth curving to body tapering upwards; fine spiral thread around mouth (now mostly lost). Fabric 2.
P.H. 3.2, diam. rim 3.7.
PN W240-W247/S352-S353 *88.20-87.65.

625   Portions of 2 necks; conical with spiral thread. Fabric 2.
P.H. 3.5 and 4, est. diam. rim 6, Th. 1.
BE-B 1966 E24-E25.5/N4-N6 *96.80-96.40.

626   Probably portion of shoulder; fine spiral thread. Fabric 2; dirty fire scum.
P.H. 5.5, Th. 0.1.
Chance find.

627   Slightly curved section of bottle(?) body; two applied spiral(?) threads. Fabric 2.
Max. dim. 2.5, Th. 0.15–0.25.
Syn MH 1963 E60-E80/N5-N10, upper fill *98.50-97.75.

628   Slightly curved portion of body (of bottle?); applied thread. Fabric 1; iridescent, dirty fire scum.
P.H. 6, Th. 0.1.
BS E 1, 1967 E6-E8/S0-S3 *97.05.

629   *Pl. 27.* Slightly curved section of vessel; two applied threads crossing one another. Probably fabric 1; dirty fire scum.
Max. dim. 5, Th. 0.15–0.2.
BS E 13, 1962 E76-E79/S2.5-S3.5 *96.40.

630   Small, slightly curved portion of vessel; two applied threads. Fabric 2; heavily corroded.
Max. dim. 2.5, Th. 0.1.
E of Syn Fc 1963 E116.8-E118/N4-N6 *97.25-96.25.

93. Crowfoot-Harden, 205, pl. 30:46, citing J. E. Quibell, *Excavations at Saqarra: 1905–1906* I (Cairo 1907) 30, pl. 34 ("late levels; Byzantine?").

631 *Pl. 15*. Portion of neck(?) of bottle(?); applied red spiral thread. Fabric 2?
Max. dim. 1.5.
B 1959, *97.40-97.00.

632 Applied blue thread of which an L-shaped section is preserved. Fabric 2.
Max. dim. 4.
BS E 12, 1964 E73-E74/S1-S2.5 *97.50-96.50.

Vessels with a Heavy Thread below Rim

A series of rim fragments of relatively thick glass have a heavy, single thread applied below the edge. The vessel walls show different angles of inclination, from about 45° to 75°. With the exception of one rim, their estimated diameters are 8 to 9 cm. It is as yet not possible to reconstruct the original vessel or vessels to which the rims belonged. We may assume, however, that all come from vessels of similar proportions and shape, perhaps from bottles with globular body, cylindrical neck, and flaring rim.[94] Although most of the rims are unstratified they seem to be datable to the Early Byzantine period although some might be of late Roman Imperial date.

The only securely dated piece (**633**) comes from BS E 19; it belongs to the few glass finds from this unit which include the remains of a goblet (type 1d) and four beaker lamps of type 4 (*BASOR* 174, 46). The other rim sections either come from upper, not safely datable levels at HoB (**634–636**), or their origin is unknown (**637–639**). Of these pieces, only one (**636**) has a smaller diameter and, therefore, might conceivably have belonged to the upper part of a more narrow bottle neck; with it were found other vessel fragments, the date of which cannot be determined as they are nondescript.

Other rim fragments of unknown provenance, including a series of specimens left uncatalogued, appear to be closely related to **633–639**. **639** shows the remains of what seems to have been a striated handle. Should this be correct one might perhaps associate rims of this particular type with late Roman Imperial bottles (**177–184**). The whole group listed in this chapter might, therefore, be dated to the late fourth or fifth century; an almost imperceptible change of style defies a clear distinction between late fourth and early fifth century glass.

633 *Pls. 15, 27*. One fifth of rim, flaring wall. Fabric 2; black scum, iridescent, corroded.
Est. diam. 8, max. dim. 4.6, Th. to 0.4.
BS E 19, 1963 E114-E116/S2-S4.

634 *Pl. 27*. One fifth of rim, flaring wall. Fabric 1.
Est. diam. 9, max. dim. 4, Th. to 0.4.
HoB 1962 E0/S115 to *102.00.

635 One sixth of rim, conical wall. Fabric 2; black scum.
Est. diam. 8, P.H. 3.8, Th. 0.25.
HoB 1961 no grid to *102.00.

636 One half of rim, flaring wall. Fabric 2; black scum, iridescent.
Est. diam. 5.8, max. dim. 5.5, Th. 0.2.
MTE 1964 E60-E65/S150-S155 *110.00-109.50.

637 One fifth of rim, flaring wall. Below heavy thread a lighter thread; object well made. Fabric 3; black scum.
Est. diam. 8, P.H. 3.3, Th. 0.3.
Chance find 1961.

638 One half of rim; at int. a tooled groove along where outer thread is applied; object well made. Fabric 2.
Est. diam. 8, P.H. 3.3, Th. 0.3.
Chance find 1961.

639 *Pl. 27*. Almost whole rim with section of upper neck, wall conical to flaring; remains of upper handle(?) with faint striations. Fabric 1.
Diam. 8.
Chance find 1961.

Vessels with Inlaid Thread Decoration

Only two fragments show thread decoration flush with the surface. Because of the rarity of this type in Sardis they could very well be imports. They belong to the relatively numerous group of vessels with sunken, often combed thread decoration that are associated with Islamic glassmaking.[95] There is no doubt, however, that at least one of the pieces is of Early Byzantine date, indicating that thread decoration marvered into the surface must have been made in the Near East prior to the Islamic period.[96]

The dated sherd, **640**, is very small; the shape of the original vessel cannot be determined. Judging by

94. For bottles and jugs with heavy thread below rim cf. Isings, nos. 102, 120ff. For a bottle of the shape described cf. Barag, "Glass Vessels," pl. 9. For similar bottle necks with applied thread cf. the finds from Jeleme, 2nd half 4th C.: Goldstein, pl. 13. Identical neck fragments from Italy with heavy thread below the rim cf. Fingerlein et al., fig. 13:1.
95. Cf. for example, Lamm, *Mittelalterl. Gläser*, pls. 29–32; the objects listed in this volume are mostly of 9th–11th C. date. Riis (supra n. 52) 62ff.
96. Cf. a series of late and post-Roman vessels with combed thread decoration that is marvered into the surface: Saldern et al., *Slg. Oppenländer*, no. 388 (listing related pieces).

the level of BE-N where it was found it dates from a period prior to the destruction of Sardis in A.D. 616 (*BASOR* 187, 57); among the other glass finds from this unit are fragments of vessels of typically Early Byzantine form as well as tesserae.

The other fragment (**641**), of blue glass with white threads, is also too small to allow a reconstruction. It comes from an area occupied by a "Roman funerary precinct" the graves of which seem to date from the fourth or fifth century (*BASOR* 191, 10ff.). However, the object might also be an Early Byzantine intrusion.

640 *Pl. 15*. Rim fragment. Greenish; yellow threads marvered into the surface (spiral pattern?).
Max. dim. 1, Th. 0.1.
BE-N 1966 E25-E28/N92-N98 *96.60.

641 *Pl. 15*. Body fragment. Blue; white threads marvered into the surface in spiral pattern. Corroded, decolorized.
Max. dim. 3, Th. 0.25.
PN 1967 W272.1-W277/S326-S329 to *87.59.

642 *Pl. 15*. G59.49a:1885. Apparently bottom section of greenish vessel with inlaid red stripes. Fabric 2?
2.4 by 1.6.
BSH, test pit in middle of the hall, N16-N19.0/W6.50-W10.50 *96.60-92.60.

VESSELS WITH MOLD-BLOWN RIBBED DECORATION

Mold-blown—more correctly pattern-molded—glass is represented in Sardis by ware with the most simple and most common motif, namely ribbing. This pattern was achieved by inflating a paraison in a ribbed mold and, after removal, expanding it so that the gather could be twisted to create spiral ribs. Mold-blown and pattern-molded glass with ribbing is frequent in late Roman[97] and Islamic times.[98] Recent finds of Sassanian glass contemporary with late Roman material also include ribbed vessels although they are different in appearance.[99]

In Sardis, pattern molding occurs predominantly on two vessel forms: bowl lamps (**237–273**) and, more frequently, bottles with or without constriction (**476–487, 519–523**). Most of the sherds listed below seem to represent ovoid bottles with necks that may have tapered upward (**647–649**); a few conical necks (**650** and **651**) could also have belonged to this series.

One fragment (**653**) was part of a bottle with bulbous body and wide neck while another (**652**) represents a cylindrical bottle. The variety of vessel shapes

with mold-blown decoration must have been greater, however, than these and a few other sherds seem to indicate (**654–656**).

Bottles with Inner Constriction

Bottles with inner constriction having a diaphragm at the lower neck were made in plain (cf. **519–523**) and ribbed ware. The largest preserved ribbed fragment (**643**) is 8 cm. high; this height seems to represent approximately one third of the body and one third or one half of the neck. Thus the height of the complete bottle was about 20 cm., a size in keeping with that of the plain bottles with constriction. The body was ovoid rather than oval (the diameter of the shoulder is larger than that of the central body); the base of such a bottle was no doubt concave while the neck either had a plain or folded lip. The ribbed pattern is most strongly pronounced at the shoulder (where the gather was not expanded), showing the actual relief of the ribs in the mold. The ridge sometimes visible on the top of the shoulder was caused by the insertion of the gather into the mold.

643 was found at the west side of BS E 13 which was particularly rich in glass and contained almost all vessel types found in Sardis, including remains of plain bottles with constriction (**519–523**, esp. **521**). A similar fragment (**644**) comes from an area at HoB where Early Byzantine material, including goblets, was discovered (*BASOR* 157, 26; *AJA* 66, 11, n.35). A smaller fragment (**645**) is from the area east of Syn Fc (*BASOR* 174, 45) where the only section of a plain bottle with constriction not found in BS E 13 was discovered.

The last of the neck-and-shoulder sections was excavated in the bath at PN (**646**) together with sherds of a knobbed goblet, a bowl lamp of type 2, and plain bottles (**481** and **490**); according to the numerous coins found in the bath, the fragments seem to belong to a period after the rebuilding of the establishment, perhaps under Theodosius II (A.D. 408–450), i.e. the fifth century or slightly later.

97. For examples of ribbed vessels of late and post-Roman date cf. Barag, "Glass Vessels," passim. Saldern et al., *Slg. Oppenländer*, nos. 491ff. For jugs (type Isings, no. 120) from Kfar Dikhrin cf. Rahmani (supra n. 77) 52–53, fig. 2:6–7, pl. 15:F–G.
98. Lamm, *Mittelalterl. Gläser*, pl. 8. For an 11th–12th C. cf. Davidson, *Corinth* XII no. 812, fig. 18.
99. Negro Ponzi (supra n. 76) 342ff. (with bibliographic ref. to mid- and late-Roman Imperial vessels with mold-blown ribs). Cf. also Puttrich-Reignard, 26ff. For Western pattern-molded and ribbed glass of the 5th–7th C. cf. esp. F. Rademacher, "Fränkische Gläser aus dem Rheinland," *BonnJbb* 147 (1942) 298ff.

643 *Pls. 15, 28.* Rounded shoulder, neck tapering upward; at int. of lower neck a constriction with diaphragm; mended. Body at shoulder vertically ribbed, with pronounced grooves at shoulder; ribs on body running to upper l. Fabric 2, thick.
P.H. 8, diam. lower neck 3.5, Th. 0.2.
BS E 13, 1963.

644 G59.5:1189. Two shoulder fragments. Similar to **643**. HoB unit 11, fill *100.00-99.80.
AJA 66, 1, n.35, pl. 7:19.

645 G63.13:5309. Fabric 2; black scum. Similar to **643**. Max. dim. 4.5.
Syn Porch E117/N0-S0 *98.00-97.75.

646 G61.21:3773. Pinched(?) depressions around shoulder from which begin ribs that are almost unrecognizable; cylindrical lower neck. Pale olive (fabric 3?).
P.H. 10, est. diam. bottle ca. 14, diam. neck 3.2.
PN W265/S360 *88.45-88.33. Found with **481** and **490**.

Bottles with Ovoid Body

A group of bottles with ovoid body and probably conical neck must have been identical or very similar to plain bottles with ovoid body (**476–487**). Although only one complete body was found (**647**)—various other sherds were too small to allow a reconstruction —a number of conical necks seem to have belonged to this vessel type. According to the evidence of the preserved body and the necks, such a bottle may have measured about 12 to 15 cm.; it had a concave base, an ovoid body and a conical neck with plain rim. The fabrics most frequently used were fabric 1 and a light yellowish green glass.

The complete body, unfortunately without neck (**647**), comes from an Early Byzantine level of the sixth or early seventh century (*BASOR* 157, 35–38; *AJA* 66, 11, no. 15). A bottle neck (**648**) which no doubt belonged to a vessel such as **647** was part of a large group of glass fragments, of which two thirds represent windows and one third vessels of typically Early Byzantine style, found in a drain at BS E 13. As the diameter of the rim is particularly large, ca. 7 cm., the original vessel must have been larger than **647**. A similar neck section (**649**) was excavated together with a series of about 75 Early Byzantine vessel fragments, nondescript in appearance, all made in fabric 1; they were found in Syn Porch where many coins from Constantine I to Justinian came to light (*BASOR* 174, 47). In addition, this room contained quantities of fragments of goblets, lamps etc. as well as pieces of cullet and glass cake.

Another neck (**650**) is datable within fairly narrow limits: it comes from the hemispherical recess at the southwest corner of Pa where a gold tremissis of Maurice (A.D. 582–602) and nine coins of Constans II (641–651) were discovered (*Sardis* M1 [1971] no. 547 and p. 155, Hoard II).

A neck with an exceptionally large diameter is listed here although it may have been part of another bottle type (**651**); found with a footed bowl (**382**), its provenance is unknown.

647 *Pls. 15, 28.* G59.72:2218. Ovoid body with concave base, fracture at lower conical(?) neck; pattern-molded ribs running to upper l. Yellowish–green (honey-colored); heavy scum.
P.H. 6.7; diam. shoulder 8.7, of base 6.
W of B W51-W52/N4-N11 *97.00-96.00.
AJA 66, 11, no. 15, pls. 7:18, 10:15.

648 Conical plain rim with ribs running to upper l. Fabric 1.
P.H. 4, est. diam. rim ca. 7.
BS E 13, 1962 in drain, *95.00.

649 Similar to **648**.
Syn Porch 1963 E117-E119/N5-N6.70 *96.75.

650 *Pl. 28.* One half of rim and upper, conical neck; ribs running almost horizontally to upper l. Fabric 1, eggshell.
P.H. 2.5, est. diam. 9.
Pa 1966 E33.60-E37.25/N16.5-N19 to *96.80.

651 Group of fragments of conical necks, some joining; ribs running to upper l. Fabric 1, eggshell.
Est. diam. 10.
BS 1963, exact findspot unknown.

Cylindrical Bottles

Only one fragment (**652**) appears to belong to a cylindrical bottle, but one or more of the ribbed sherds too small to allow reconstruction may have been part of such a bottle. **652**, which may be assigned to the first phase of Byzantine occupation at AcT, was associated with coins of Maurice (A.D. 582–602) although a later dating is possible (*BASOR* 170, 31–33; *Sardis* M1 [1971] no. 651).

652 Lower portion with flat base; pattern-molded with ribs to upper r. Fabric 1.
P.H. 3.6, diam. 4.5, Th. 0.2.
AcT 1962 E10-E16/N17-N20, level Ib, unit 7, *403.57-403.37.

Globular Bottles with Wide Neck

A bottle with wide neck and perhaps an oval body is apparently represented by **653**; its reconstruction is practically impossible as neither the upper neck nor the lower body are preserved. However, the original piece may have been similar to plain bottles with wide neck (cf. **509–516**).[100] The fragment comes from BE-E where window fragments of fabric 1 and a marble head reworked in the fifth or sixth century were found, suggesting its Early Byzantine date (*BASOR* 177, 25; *Sardis* R2 [1978] no. 94 with earlier references).

653 *Pl. 15.* Small fragment of (upper part?) of vessel slightly tapering upwards, with rim(?) bent outward; ribs run vertically. Fabric 1.
Max. dim. 3.8, Th. 0.1.
BE-E 1964 E11-E13/N27-N32 *97.00-96.50.

Miscellaneous Vessels

The following fragments were part of vessels the exact shape of which must remain unknown until more complete parallel pieces are excavated. The fragment of a ribbed vessel (**654**) is certainly of Early Byzantine date (*BASOR* 154, 16–18); it may represent the type of wide-necked bottle referred to in the foregoing (cf. **653**). The pattern-molded, ribbed, Middle Byzantine vessels found in Corinth to which this piece has been compared were certainly made the same way[101] and the vessel shapes may have been very similar.[102]

While a series of pattern-molded fragments from two bottles(?) and one bowl (**655**), found at BSH inside the south apse, may belong to the late Roman or Early Byzantine periods (*BASOR* 154, 13–16), another group of fragments (**656**) from the northeast corner of BE-S probably belongs to the early fifth century (*BASOR* 162, 40–43). The latter group was found with a multitude of typically Early Byzantine vessel fragments: threaded bottle necks (cf. **607–621**), lamp handles, goblet stems, concave bases of vessels, and plain vessel rims. There is a strong probability that at least some of the fragments (**656**) come from cylindrical bottles (cf. **652**).

654 G58.47:612. "Short neck going without break into vertically fluted (ribbed) body" (recorder). Fabric 1.
BS W 1, *97.30-96.60.
Hanfmann, *JGS*, 53.

655 G58.74:879. Two joined fragments of glass bowl with horizontal ribs. Silvery gray green patina (according to excavation records).

P.W. 7.0, P.L. 8.3.
BSH.
Hanfmann, *JGS* 53, n.16.

656 Vessel fragments; pattern-molded ribs. Fabric 1?
BE-S 1960 E33-E35/N38-N42 *97.00-96.25.

MILLEFIORI GLASS

At Sardis, finds of millefiori glass are relatively rare. While millefiori and mosaic glass was comparatively frequent in late Hellenistic and early Roman times, its manufacture became the exception rather than the rule in the late and post-Roman period.[103] It was not before the ninth century that Islamic glassmakers rejuvenated this technique to make vessels as well as small panels for wall decoration.[104] The Sardis finds include ten fragments, eight of which were part of flat plaques and one or two belonged to vessels. They all seem to be imports; if they had been made locally, more objects in this technique would have been found. The plaques may have been used as wall decoration embedded in plaster.

The glass is green with white circular canes, perhaps an indication that all of the pieces have a common provenance; three fragments of plaques have yellow canes (**664–666**). As this color combination and a pattern of circular canes are not only frequent in Roman Imperial glass but also occur, in slightly different form, in ninth century Islamic glass, there seems to be no way to date the Sardis finds on stylistic or technical grounds alone. According to the archaeo-

100. For a squat, globular bottle with wide neck cf. G. F. Bass and F. H. van Doorninck, Jr., "A Fourth-Century Shipwreck at Yassi Ada," *AJA* 75 (1971) 37, pl. 3:33. Cf. very similar, wide-necked bottles with vertical ribbing: Riis (supra n. 52) 51, fig. 130 (Islamic).
101. Davidson, *Corinth* XII nos. 780–784.
102. Cf. ibid., esp. no. 795.
103. For Hellenistic millefiori glass cf. A. Oliver, Jr., "Millefiori Glass in Classical Antiquity," *JGS* 10 (1968) 48–69. For early Roman millefiori vessels and plaques cf., for example, Saldern et al., *Slg. Oppenländer,* nos. 309ff. (with ref. to earlier literature).
104. Millefiori glass from Ctesiphon (ca, 5th–6th C.?): Clairmont (supra n. 59) 144, no. 5; a chunk of millefiori glass in red, turquoise-blue, white, yellow, and blue, perhaps datable to the same time, in the Metropolitan Museum, inv. no. 32.150.55. For Islamic millefiori glass, including vessels and tiles, which comes mainly from Samarra (9th C.) cf. Lamm, *Samarra,* 108–110, pls. 9, 12; *Islamische Kunst,* exhibition catalogue Museum für Islamische Kunst, Berlin (1967) no. 51, color pl. 1; cf. also the millefiori glass from Samarra in the Victoria and Albert Museum, London (inv. no. C.742-8-1922) and variously shaped inlays of clear glass. For Islamic millefiori glass without provenance cf. *Smith Coll.,* nos. 481ff.; Clairmont (supra n. 59) 144–145, no. 6.

logical data, most of the pieces appear to belong to the Early Byzantine period. If this proves to be correct, one is led to conclude that somewhere within the Byzantine realm one or more workshops continued to make millefiori glass in the Roman tradition, serving as a link between late Roman and early Islamic manufacture of this material. With the exception of **663–666** the finds come from the area of the Gymnasium.

A piece (**657**) found at the wall behind the apse of the Synagogue at the northeast corner of BE-B, may be datable to very late Roman Imperial times or to a period after A.D. 400. From the same location comes an unstratified plaque section (**658**). A section of a similar plaque (**659**), from a dump at BE-B, is also unstratified. Of approximately equal size and thickness is a fragment from an area east of Syn (**660**). Other sections of the same millefiori glass were excavated in BS E 18 (**661**; the shop is dated to the sixth century), together with pieces of blue tesserae or cullet (**711**), and east of Syn Fc (**662**) where much Early Byzantine glass came to light.

Fragment **663** perhaps formed part of the rim of a hemispherical bowl, a form particularly popular in late Hellenistic and early Roman times. Although not stratified, this fragment appears to be later: the green color and the circular canes link it with millefiori glass of Early Byzantine date found in Sardis.

Three additional fragments of plaques (**664–666**) from the Synagogue could be either Roman or Early Byzantine.

657 *Pl. 16.* Part of plaque (wall decoration?). Green with white circular canes.
Max. dim. 2.5, Th. 1.
Syn 1966 E31.50/N8.00, on wall behind apse.

658 G66.4:7019. Irregular section of plaque. Green with white circular canes.
Max. dim. 1.7, Th. 1.
Syn, on apse wall, unstratified.

659 Underside of plaque has irregular surface and looks as if lumps were added. Probably green with white circular canes; dirty, decolorized.
Max. dim. 4, Th. 0.9.
BE-B 1966, in dump ca. E30/N1 (probably from Syn).

660 Irregular section of plaque. Green with white circular canes.
2.3 by 2.5, Th. 1.1.
E of Syn 1968 E117.5-E120/N23-N27 *97.50.97.12.

661 Small fractured section of a plaque. Similar to **658** and **659**.
Max. dim. 2.5, Th. 1.
BS E 18, 1963 E105-E109/S0-S3 *97.50-97.00.

662 *Pl. 16.* G63.11:5302. Underside of rim(?) section of plaque, very irregular and lumpy (probably used as wall decoration). Green with irregular white circular canes; decolorized; mended.
9.5 by 4.5, Th. 0.6–0.9.
E of Syn E117-E121/N2-N4 *96.75-96.25.

663 *Pl. 16.* Small rim section and another fragment of shallow (or almost hemispherical?) bowl. Green with white circular canes.
Max. dim. 2.5, Th. 0.3.
HoB 1962 W23/S93, brown earth.

664 *Pl. 28.* Probably part of inlay plaque. Pattern on back side irregular and "squashed." Blue with yellow(?) circular canes; corroded.
Max. dim. 6.3, Th. 0.8–0.9.
Syn 1965, with fallen skoutlosis fragments, unstratified.

665 *Pl. 28.* Part of inlay plaque. Pattern on back side unrecognizable because of weathering. Blue with yellow irregular canes; back side heavily irridescent.
Max. dim. 5, Th. 0.8.
Syn 1965, with fallen skoutlosis fragments, unstratified.

666 *Pl. 28.* Portion of plaque, perhaps from same piece as **665**.
Max. dim. 3.7, Th. 0.13.

STAMPS

Eight glass stamps with Christian and Jewish symbols were either used as merchandise marks, weights, tokens, tickets, amulets, or they were attached to vessels (cf. **673**).[105] They are circular, having an average diameter of 2 to 3 cm., and show on their obverse a symbol—for example a cross or a menorah, often combined with letters—stamped onto the surface while the glass was still hot.

Stamps dating from the Early Byzantine period have been found predominantly in Egypt, Syria, Crete, and Cyprus while finds from Asia Minor seem to have been the exception.[106] Most of them are issues of the reigns of Justin I (A.D. 518–527) and Justinian (527–565), a period to which the stamps found in Sardis can also be assigned.[107] Whether or not they

105. For their use cf. Harden, *Karanis,* 297ff.
106. G. Schlumberger, "Poids de verre, étalons monétiformes," *REG* 8 (1895) 59ff. Cf. also P. Balog, "Poids monétaires en verre byzantino-arabes," *RBN* 54 (1958) 127–137.
107. G. V. Gentili, "Les poids monétaires en verre byzantins, byzanto-arabes et arabes provenant des fouilles de la villa romaine de Piazza Armerina," *Annales du 4e Congrès des "Journées Internationales du Verre," Ravenne-Venise, 1967* (Liège [1969]) 133ff.; cf. also Philippe, 36ff. (with bibliographic ref.).

are of local origin cannot be decided at present. The glass, of various shades—particularly aquamarine and pale green—could very well be linked to the fabrics manufactured in Sardis, namely fabrics 1 and 2. All of the finds come from the Gymnasium area, particularly from RT, Fc, and Pa. At least four of them are Christian while at least one (**674**) is Jewish.

The first (**667**) probably has the chi-rho sign; it comes from east of Fc where coins of the late fourth to sixth century as well as much Byzantine glass and tesserae were found. Another stamp (**668**) shows a cross and the letters NPEA; it comes from BS E 14 where it was found together with two other stamps (**669** and **670**; *BASOR* 170, 50). One of these pieces has a cross-like symbol with its bars terminating in double lines and V-shaped motifs; the design on the other is unidentifiable.

A stamp with cross and letters (**671**) comes from RTW where it was found with fragments of Byzantine window panes.[108] A stamp from Pa-W no doubt belongs to a fifth to sixth century level (**672**); it bears a monogram in the form of a large N, each bar terminating in letters.[109] A stamp with a C and an inverted V, found at Syn Fc (**673**) is exceptional in that it is attached to the rim of a conical bowl or a bottle with conical neck. Stratigraphic evidence suggests a date of perhaps the late fourth or fifth century. The last stamp listed here (**674**) shows a menorah and the letters CN.[110] It comes from Pa-S; window glass of Early Byzantine date was found in its immediate vicinity.

667 *Pls. 16, 28.* G63.7:5249. Circular with bulgy rim. Design (very soft): cross with X(?) and P. Greenish?
Diam. 2.4.
E of Syn E119.5-E121/N3.5-N4.5 *97.00-96.75.

668 *Pls. 16, 28.* G62.5:4163. Circular with rim. Design: cross with letters N P E A (Andrea?; cf. Schlumberger, supra n.106). Greenish?
Diam. 2.1.
BS E 14, E83/S2 *97.10. Found with **670**.
Philippe, fig. 15.

669 *Pl. 28.* G62.4:4162. Circular with bulgy rim. Design: cross with horizontal bar terminating in perpendicular double lines, the vertical bar ending in V-shaped angles. Fabric 1.
Diam. 2.5.
BS E 14 E84/S2.30 *97.30-97.10.

670 G62.6:4164. Circular with rim. Design unidentifiable. Bluish-green (similar to fabric 1).
Diam. 1.8.
BS E 14 E83/S2 *97.10. Found with stamp **668**.

671 *Pl. 16.* G62.9:4340. Circular with rim. Design: cross and H V (?) K A. Pale yellowish-green.
Diam. 1.8.
RTW E32/S29 *96.00-95.50.

672 *Pl. 28.* G67.2:7287. Circular with small section of vessel rim. Design: monogram with large N incorporating A (and O, Y?). Greenish?
Diam. 1.5.
Pa-W E33.80-E36.80/N92-N95.70 *97.70-97.50.

673 *Pl. 28.* G67.4:7315. One-quarter section of rim and upper part of conical neck (of bowl?). Circular stamp attached having bulgy edge. Design: monogram-like C with inverted V. Green tint, eggshell.
Est. diam. 6, max. dim. 3.5, diam. stamp 1.6.
Syn Fc E113-E114.55/N4-N10 *96.45-96.00.

674 *Pl. 16.* G67.5:7401. Circular-irregular with rim. Stamped menorah and the letters: C N. Greenish (fabric 2?).
Diam. 1.7.
Pa-S E73.14/N27.75 *96.56.
BASOR 191, 28, fig. 20.

RODS

A few pins or rods were found in Early Byzantine levels. They may have been used as stirring rods (cf. **224–228**). One rod (**675**), discovered together with a metal rod, comes from Syn Fc at a level after the reconstruction of about A.D. 400. **676**, found with a conical bottle neck (**503**), was among the finds from BS E 16 where other Early Byzantine material came to light. The third rod (**677**) is equally well dated as it was found in BS E 13.

675 *Pl. 16.* Section of pin. Greenish (fabric 2?); heavy decomposition. Found with metal rod.
P.L. 7.7, diam. 0.8.
Syn Fc 1963 E105-E110/N1.20-N3.00 *97.50-97.00.

676 *Pl. 16.* Portion of slightly tapering rod with rounded end. Heavy silver iridescence.
P.H. 5.2.
BS E 16, 1963 to *96.40. Found with **503**.

---

108. For a similar stamp cf. Schlumberger (supra n. 106) no. 15.
109. For other stamps with large N cf. ibid., no. 34; M. Jungfleisch, "Les dénéraux et estampilles byzantins en verre de la Collection Froehner," *BIE* 14 (1932) 235, 244, 246.
110. For a menorah on a stamp attachment of a vessel cf. D. Barag, "A Jewish Applied Glass Medallion," Museum Haaretz, *Bulletin* 11 (June 1969) 39–42 (perhaps 4th C.).

677 *Pl. 16*. Lower portion of "double" stick i.e. 2 sticks or rods joined together. The end was roughly knocked off at manufacture. Greenish (fabric 2?).
P.H. 4.5, W. 1.7.
BS E 13, 1962 E77-E80/S1-S3 *96.70-96.30.

BRACELETS

Sections of only two bracelets were found in Sardis that can be assigned to the Early Byzantine period (cf. **232** and **233** and **738–779**). Of fabric 3, **678** comes from Syn Fc, belonging certainly to a time within the Early Byzantine period. The other (**679**) was found in BS W 14 together with remains of Byzantine vessels.

678 Section of bracelet. Olive-green (fabric 3?).
Syn Fc 1964 E108-E110/N4 *94.50.

679 *Pl. 16*. One third of bracelet, with biconical cross section. Blue; dirty, scratches on shiny surface.
Est. diam. 5.5, W. 1.
BS W 14, 1966 W65.70-W68.50/S3.10-S4.80 *97.00-96.30.

WINDOW GLASS

Great quantities of window glass were found in Sardis. In fact, the total amount of sherds surpasses, in sheer weight, the total amount of fragments of glass vessels. Window glass sherds probably represent thousands of panes. Most of them come from the Byzantine Shops and the areas adjacent to the Synagogue and the Gymnasium.

The dating of practically all of the panes recorded is identical to that of the glass vessels, i.e. they were made between the early fifth and the early seventh centuries.[111] The glass appears to have been manufactured in the same workshops since identical fabrics—particularly fabric 1—were used both for flat and hollow wares. Two thirds to four fifths of the total number of finds are of aquamarine (fabric 1) glass; the balance comes in various shades of green, most of which is not unlike the bottle olive glass (fabric 3). One of the few fragments of clear glass (**680**) was found in BS E 12, which may suggest that finds of clear glass do not necessarily exclude an Early Byzantine dating.

The fragments have an average thickness of 0.2 cm.; a thickness of 0.35 cm. occurs very rarely. Panes 0.15 or 0.4 to 0.5 cm. thick, also of fabrics 1 and 3, are the exception. Apart from weathering, either side of a given pane is glossy.

The flat glass was made by blowing a cylinder, cut-ting it open and placing it on a flat surface. A number of fragments are preserved that show a smoothly rounded rim, either straight or slightly curved, which is evidence of this type of manufacture.[112] After cooling, the entire piece was used as a window pane, or it was cut into smaller sections. Most panes seem to belong to what D. B. Harden called the "double-glossy" type; however, some fragments, for example from BS E 12 are of the "glossy-mat" type.[113]

The largest section from a window pane found in Sardis has a length of 32 cm. (**681**).[114] Thus, panes as large as 30 by 40 cm. must have been made. The windows in BS were about 83 to 102 cm. in width while one grill measured 72.5 by 47.5 cm. It is as yet impossible to say whether fragments with slightly curving rims are indicative of arched windows.[115] We are inclined to believe that those with curved edges—only very few of which have been recorded—were used as large circular or irregularly shaped panes.

"Bull's eye" panes, made by the crown glass method—rotating the pontil until the gather flattens out and then removing it from the rod, leaving a pontil mark at the center—have not been found in an Early Byzantine context in Sardis.[116]

111. For contemporary window glass from Jerash made of similar fabrics and often showing identical weathering cf. Baur-Kraeling, 546; the panes have the same thickness as the Sardis panes, and the edges are also either straight or slightly curved. Cf. also D. B. Harden, "Roman Window Panes from Jerash and Later Parallels," *Iraq* 6 (1939) 91. Panes with identical weathering from Auja Hafir in the Ashmolean Museum, Oxford. "Nearly 300 fragments (of window glass)" from Shavei Zion are datable to the 5th–6th C.: Barag, *Shavei Zion,* 69–70. Cf. also Saller, *Moses,* 64–65, 314. See *Pl. 16* for sample fragments of panes from Sardis.

112. For recent studies on window glass cf. esp. T. E. Haevernick, P. Hahn-Weinheimer," Untersuchungen römischer Fenstergläser," *Saalburg Jahrbuch* 14 (1955) 65ff. D. B. Harden, "New Light on Roman and Medieval Window Glass," *Munich Congress,* VIII.2. Idem, "Domestic Window Glass: Roman, Saxon and Medieval," in E. M. Jope ed., *Studies in Building History: Essays in Recognition of the Work of B. H. St. J. O'Neil* (London 1961) 39–63. G. C. Boon, "Roman Window Glass from Wales," *JGS* 8 (1966) 41–45 (with remarks on manufacturing techniques particularly of the "glossy-mat" type). R. Cramp, "Glass Finds from the Anglo-Saxon Monastery of Monkwearmouth and Jarrow," *London Congress,* 16–19 (late 7th–mid-9th C.). W. Haberey, "Funde aus Glas," in *Ein Burgus in Froitzheim, Kreis Düren, Beiträge zur Archäologie des romischen Rheinlands* (Düsseldorf 1968) 87–88.

113. Boon (supra n. 112) 42.

114. According to G.M.A. Hanfmann (*JGS*, 52), fragments up to 40 cm. in W. were excavated.

115. Hanfmann, *JGS,* 52.

116. Crown glass appears to have been found in various contexts in Churches E and EA datable to Middle Byzantine times—crown glass seems to have been introduced in the 4th C. Some of the panes from Jerash were made that way: Baur-Kraeling, 546. Cf.

The panes were installed in windows with the help of lead strips having an average thickness of 0.3 cm. and bent twice at right angles to form a Z-like profile. The two terminal sections of such a strip have a length of about 2.2 to 2.5 cm. while the middle section is about 1.5 cm. long. Strips of this type appear to have been fastened by plaster to the window casing, and the glass panes were pushed against the strips and held in place by some kind of mechanical device (or plaster). Among the well-preserved strip sections is one from BS E 6 (*Pl. 16*.), datable before the early seventh century; other lumps of lead, perhaps molten strips, were found at the same location and in other shops (E 7 etc.).[117] Leading also comes from in front of BS E 8. It is difficult to explain why so much window glass was found in and near the Byzantine Shops; the number of windows in the shops themselves cannot account for the large quantity of panes. On the other hand, if the panes found were part of the windows of the Synagogue, an equal number of sherds should have been found around the entire structure. Thus, it seems likely that at least a portion of the fragments of flat glass from the Byzantine shops belonged to the stock of these establishments.

A listing of sample pieces should suffice to demonstrate the full range of characteristics of the window glass under discussion—sections with edges, color, size etc. Numerous fragments came from BS E 12 at a level of *98.25–98.00 where they were found together with goblet stems, bottle necks with spiral thread (cf. **607–621**), remains of lamps, etc. About fifty pounds of fragments were collected in 1967 in BS E 6 at *97.48 to *96.90 (*BASOR* 191, 18, 20) where coins of Arcadius and Honorius (A.D. 395–408) and Heraclius (A.D. 610–616) came to light; a series of lead sections were excavated at the same spot.

680   Clear fragment with green tint. Th. 0.3.
BS E 12, 1964 E70-E75/S1-S3 *98.25-98.00.

681   G58.17A,B:469; G58.18:470. Fragments. Fabric 1 and green.
Max. dim. 32.
BS W 1, W3.25/S0-S3, level II.
Hanfmann, *JGS*, 52, fig. 1, left.

Sample of Window Fragment Findspots

A large proportion of all the fragments found come from these locations. A selection of fragments is illustrated on *Pl. 16*.

682   BE-A 1967 E6-E10/N0-N2 *97.35.

683   BS E 1, 1967 E7-E10/S0-S6 *97.87.

684   BS E 1, 1967 E8.5-E10/S0-S6 *97.30.

685   BS E 4, 1967 E30.50-E33/S0-S5 *98.20-97.20.

686   BS E 6, 1967 E41-E44/S0-S4 *97.81-97.31.

687   BS E 7, 1967 E49.73/S3.25 *97.32.

688   BS E 12, 1964 E70-E75/S1-S3 *98.25-98.00.

689   W of B 1966 W60.5-W66.5/S3-S6 down to *96.60.

690   W of B 1966 W62.25-W66.25/N0.35-S2.25 to *96.35.

691   Pa-S 1967 E86-E90/N20.70-N27, removing fallen wall, *97.20-96.20.

692   Pa-S 1965 E72-E74/N20.8-N23.5 *96.40-95.00.

693   BE-A 1966 E6-E10/N0-N9.80 ca. *98.70-97.20.

694   RT 1962 E32-E35/S29-S31, S of colonnade foundation, *98.00-96.53.

695   RT 1961 E10-E13.1/S5.30-S9.00, fill above mosaic pavement, *96.72-96.40.

696   G61.14:3412. Largest fragment found after ca. 1960. Max. dim. 16.
RT E5-E9/S7, on top of marble floor *96.70.

697   PN 1961 W255/S360 to ca. *88.78.

698   PN 1963 W228-W232/S349-S350 *88.40-87.90.

699   Ca. 2 pounds of fragments.
CG 1959, northernmost arch, ca. *96.70.

TESSERAE

Mosaic cubes or tesserae of glass were found in great quantities together with other Early Byzantine material. They were used as wall decoration. Accord-

---

also Crowfoot, *Samaria*, 420–421, and Harden, *ArchJ*, 83. R. Chambon, "L'évolution des procédés de fabrication manuelle du verre à vitres du dixième siècle à nos jours," *Washington Congress*, 165–178.
117. For H-shaped leading used in the West cf. Cramp (supra n. 112) 16. Circular crown glass panes were set in plaster; cf. examples from Jerash, 6th–7th C.: Harden, *ArchJ*, 83 and pl. 6:A, G. Another method of holding the pane in place was to encase it in plaster: cf. Saller, *Moses*, 64–65, figs. 12–13, pl. 130, figs. 1–2.

ing to archaeological evidence they do not appear to have been employed to cover a whole wall but rather seem to have served as decorative friezes, bands, and dividing lines within a large frescoed section (cf. BE-N; *BASOR* 187, 56).[118] Tesserae found in Sardis that cannot be closely associated with Early Byzantine material may be of late Roman date; at least one, with gold foil, appears to date before A.D. 400.

The tesserae, knocked off with a hammer from glass "cakes," were made in various colors. The majority is of a light, translucent green glass; blue, manufactured at least in three different shades, appears to have been less frequent. In addition, there is opaque black, purple, milky purple, dark red, three tints of yellow, transparent yellowish-green as well as almost clear glass. The clear material, generally slightly tinted green or yellow, was used almost exclusively for sandwiched gold leaf tesserae. The cubes vary in size from ca. 1.0 by 0.4 by 0.6 cm. to 1.0 by 1.2 by 1.1 cm. with median dimensions of 1.0 by 0.7 by 0.8 cm. While the majority of the tesserae were found as single cubes, divorced from their original context, a number came to light as fragmentary panel sections still set in their bedding. The glass tesserae, occasionally combined with stone cubes, were stuck side by side into the wet mortar which has an average depth of between 2.2 and 4.0 cm. The surface of the tesserae is about 0.8 cm. above the bedding (cf. *BASOR* 206, 20).

It has not yet been possible to reconstruct a complete design or pattern. Apparently, in most cases the cubes were assembled into geometric designs. L. J. Majewski recognized in at least one section, **700**, part of a face (nose), proof that glass and stone tesserae were combined to form figurative scenes; this section has been assigned to the fifth century. Tesserae found in 1971 show that portions of the upper wall of the Synagogue were decorated with mosaics in arabesque patterns and with mosaic inscriptions (*BASOR* 206, 20).

The tesserae are generally found with the remains of Early Byzantine glass vessels of the types discussed in the foregoing sections. Particularly rich finds come from the Gymnasium (LNH, BE-H, etc.) but also from HoB, PN, and CG. They are relatively rare in the Byzantine Shops.

The abundance of finds suggests that the tesserae must have been made locally. However, as practically no remains of colored glass vessels came to light, one is led to assume that either there were special workshops producing colored glass for tesserae, or, less likely, one or more of the establishments were in-

volved in two separate operations: one for hollow and flat glass, the other (with imported cullet?) for tesserae.

One of the major finds comes from LNH (**700**). It includes about sixteen fragmentary panel sections, each showing about twenty polychrome tesserae of glass and, occasionally, of stone. Remains of a series of lamps of type 4 were discovered at the same spot. As mentioned before, this group is datable to the fifth century (L. J. Majewski). A similar find comes from BE-N (**701**); according to the level (*98.15) it is apparently datable to a period close to the destruction of the building in A.D. 616 (*BASOR* 187, 56).

Another group find stems from the center of Syn (**702**); these cubes represent at least a dozen shades of color and seem to be of the late fourth or fifth century (*BASOR* 206, 20, 37ff.).

Among the other finds is a group of tesserae still in their bedding, found in BE-H (**703**). Another large group of cubes was found in BE-A (**704**); although not associated with a clearly defined level, they no doubt belong to the Early Byzantine period. From BE-H come many tesserae found in the fill covering a floor at *96.00 (**705**); they, too, are most likely of Byzantine date. A group, including tesserae in transparent clear glass with yellow tint, was discovered at BE-B (**706**) at a relatively high level; however, this area has no doubt been disturbed since the cubes should be datable to a period prior to A.D. 616.

Tesserae in clear glass with sandwiched gold leaf, very common in mosaics covering the walls of early Christian and Byzantine churches in Rome, Ravenna, and Constantinople, are quite rare in Sardis.[119] They were made by facing the cube with gold leaf and, as a protective coat, fusing a thin layer of clear glass (0.1 cm. thick) to the top. As the manufacture of this type

118. Many post-Roman sites in the East have revealed glass tesserae which often seem not to have been included in the publications of the finds: see Harden, *ArchJ*, 83. For general remarks see Philippe, 23ff. Tesserae identical to those found in Sardis come from Jerash (5th–6th C.; Baur-Kraeling, 518), Bethany (5th C.; Saller, *Bethany*, 42–43, 326–327), Mount Nebo (5th C.; Saller, *Moses*, 211ff.), and other places.

For tesserae and chunks of opaque glass of the 4th C. and later cf. Weinberg (supra n. 49) 133. For late 6th–mid-7th C. tesserae from a factory at Torcello cf. Leciejewicz et al., ". . . Torcello nel 1961 . . ." (supra n. 51) 47; E. Tabaczynska," Glassworks on Torcello near Venice in the 7th and 8th Century; Excavations 1961 to 1962," *Bruxelles Congrès*, 238.2.

119. An ingot or cake with sandwiched gold foil used to make mosaic cubes was found in Israel: Goldstein (supra n. 49) 129 ("late Roman/Early Byzantine").

of tessera is not difficult, their infrequent occurrence in Sardis may indicate that they were imported or they might be reused late Roman Imperial mosaics. One of them (**707**) comes from BE-B where it was found together with typically Early Byzantine glass; coins from this level date from the fourth to sixth century.

Fragmentary cubes of yellowish glass as well as with gold foil were discovered east of Syn Fc (**708**), a location rich in finds of Early Byzantine glass vessels and cullet.

Another gold foil cube comes from HoB (**709**); its level (*100.90–100.70) indicates that it may be of Roman (second to third century) date. Among the glass objects found in the same year in its immediate neighborhood, although at a higher level (*102.00), are sections of an early Roman pillar-molded bowl (**33**), a wart-like attachment of another bowl (**90**), and a bottle neck with spiral thread which appears to be Early Byzantine (**612**).

Three small tessera-like objects of blue glass, either cullet or used in mosaics, come from BS E 18 (**711**) where they were found with a section of a millefiori glass plaque (**661**). The material found in this shop is datable to the mid-sixth century.

700 *Pl. 17.* Group of tesserae in plaster bedding. Ca. 16 fragmentary sections, each with ca. 20 tesserae. Glass and stone mixed. Colors of glass: yellow, green (in at least 2 shades), blue, blackish, purple, red. One cube is clear with sandwiched gold foil. Many of the cubes have porous surfaces (caused by burning: yellow turning black). According to L. J. Majewski part of a face (nose) is recognizable in the design.
Tesserae: 0.6–1; Th. of bedding ca. 3–4 (2 kinds of bedding).
LNH 1, 1968 E56-E61/N106.16-N110 *97.5-97.30.

701 Group of tesserae in plaster bedding. Glass: blue, yellow, black(?).
Tesserae: 0.6–1.
BE-N 1966 E32.7/N93.2 *98.15.
BASOR 187, 56.

702 Group of tesserae in plaster bedding. Glass: black(?), dark red, 3 shades of blue, 3 shades of green, 3 shades of yellow, greenish and yellowish, clear. Stone tesserae in white, pink, and gray marble, black basalt, yellow and brown calcareous stone; mother-of-pearl tesserae.
Glass: 0.3–1; Th. bedding 3.
Syn MH 1971 E62.3-E63.2/N12.5-N13, in depression filled with brick and rubble.
BASOR 206, 20.

703 Group of tesserae in plaster bedding. Mainly blue and green cubes.
Tesserae: 0.6–1; Th. bedding 2.2, tesserae extend 0.8 above bedding.
BE-H 1966 E5-E7/N69-N72 *96.90-95.50.

704 Group of tesserae of various colors.
BE-A 1966 E7.50-E10.50/N0.90-N9.90.

705 Group of tesserae. Translucent and opaque blue and green, mixed with stone tesserae.
BE-H 1964 E5-E10/N64-N66 to *96.00.
BASOR 177, 23.

706 Group of tesserae. Various colors, including transparent yellowish.
BE-B 1966 E18-E21/N9-N10 *98.80-98.00.

707 *Pl. 17.* Greenish–olive covered with layer of yellowish glass; sandwiched gold leaf.
1.1 by 0.9 by 0.8, Th. upper layer 0.1.
BE-B 1966 E26.50-E28.50/N7-N9 *96.80-96.40.

708 Fragments of tesserae. Yellowish; gold foil.
E of Syn Fc 1963 E119.2/N2-N5 *97.75.

709 Clear; gold leaf applied on one surface, without protective layer (lost?).
0.8 by 1.2 by 0.9.
HoB 1965 W35-W37.50/S113-S118 to *100.90-100.70.

710 Bottle green (fabric 4?), bubbly. Covered with clear layer with green tint; sandwiched gold leaf (covering layer has come off).
0.6 by 1 by 0.6.
Chance find 1964.

711 Three objects (cullet or used as tesserae). Blue, irregular.
BS E 18, 1963 E105-E109/S0-S2 *97.50-97.00.

CULLET

Cullet, or raw glass, is unassuming in appearance, but from a technological point of view there are few finds of more interest than cullet and wasters—drippings etc.—which are proof of the actual manufacture of glass at a site under investigation.[120]

120. For comparable cullet from Aphrodisias, ca. 5th–7th C. cf. R. H. Brill," The Scientific Investigation of Ancient Glass," *London Congress,* 51, fig. 2; from Corinth, late 5th–6th C. cf. Wiseman (supra n. 7) 105–106. For reports on factory sites, cullet, drippings etc. cf., for example, Tabaczynska (supra n. 118) 238.1–3; idem,

Cullet is a generic term for all kinds of glass fragments and unshaped lumps of glass left as residue in the melting pots. The cullet is broken off the interior of pots no longer usable and employed again as raw material to facilitate the melting process of a new batch of glass. Finds of cullet alone are, admittedly, no proof that glass was actually manufactured at a given location. Throughout the history of glass, cullet has been a fairly common merchandise that was shipped to localities not equipped to produce half-melted glass or regular cullet in sufficient quantities for their own production.[121]

In Early Byzantine Sardis, the numerous fragments of glass vessels and windows homogeneous in style and appearance as well as the discovery of the remains of crucibles and drippings, together with the finds of cullet, should be considered proof of the existence of glassmaking facilities in Sardis proper.

Many pieces of irregularly shaped lumps of cullet were found. They are of dark green (fabric 2?) and aquamarine fabric 1) glass; the former, because of the density of the material, is black in appearance. The lumps are small in size. They are frequently partly coated with a ceramic-like material which appears to come from the melting pots or crucibles.

For obvious reasons only a relatively small number of pieces have been recorded in this catalogue as a listing of all cullet excavated in Sardis would have been redundant. The majority of the finds come from the areas of the Synagogue, the Byzantine Shops, and BE. It is somewhat curious to note that much of this material was found near the remains of glass vessels; one would expect this to be the exception and not the rule. According to archaeological evidence, one might speculate that the manufacturing facilities were located not far from the Syn and B complex. Some of the cullet (**712** and **713**) was found in the debris of BE-H and at BE-S (*BASOR* 177, 23–25). Although certainly Byzantine, the pieces cannot be dated more precisely; in BE-S were also found sections of goblets and bowl lamps of type 2.

More cullet came to light E of Syn Fc (**714**; *BASOR* 174, 43–44), in BE-C (**715**) and in BE-B (**716**). The latter unit also contained fragments of vessels and windows (*BASOR* 187, 18). All finds in this group should be dated to the early fifth century. Lumps of cullet also come from BS (**717**), probably from E 4 to E 8 where every type of Early Byzantine glass was found (*BASOR* 191, 16ff.). Among the cullet less securely dated is a pumice-like lump from PN (**718**) that was found together with window glass and, according to G. M. A. Hanfmann, seems late or post-Roman

(*BASOR* 177, 6). A lump from HoB (**719**) is unstratified.

Cullet with ceramic matter adhering to it and fragments of melting pots or crucibles coated on the interior with a heavy glaze or a thick layer of glass should be taken as incontrovertible proof of glass manufacture in Sardis. All such objects found are recorded in the catalogue. Most of the cullet mentioned in the following undoubtedly stems from broken crucibles which, after having been in use for some time, disintegrated or were coated with such a heavy layer of glass that melting became difficult. Almost all of the objects listed here are of dark green glass appearing black in reflected light (fabrics 2 or 4). **721** was discovered in BS E 4, which contained cullet without ceramic material adhering to it as well as many vessel fragments (*BASOR* 191, 17).

At least one section of a melting pot has been identified (**722**); its inner, concave surface is lined with yellowish green, bubbly glass. It was discovered in BE-B just above the floor of the rebuilding period in the early fifth century (*BASOR* 187, 15–16). Tesserae and vessel fragments were found in the immediate neighborhood.

It is difficult to say whether the ceramic sections (**723** and **724**) with vitrified interior, which were found at MTE, were part of pots or whether they are the remains of glazed bricks that vitrified when they were exposed to intense fire (as suggested by L. J. Majewski). Unfortunately, they come from an unstratified dump that largely contained material from the late Hellenistic to the late Roman periods (*BASOR* 187, 14–15).

A series of vitrified lumps of clay (**725**), or rather glass mixed with a ceramic material and showing signs of exposure to great heat, are from BE-C. According to G. M. A. Hanfmann they were dumped, perhaps in the sixth century, to the level of the floor; at any rate they are definitely after the time of the remodeling early in the fifth century (*BASOR* 187, 19–20). Finally, a section, probably of a melting pot (**726**),

"Remarks on the Origin of the Venetian Glassmaking Center," *London Congress,* 20–23 (dating the furnace to the late 6th–mid-7th C.). A. Gasparetto," The Torcello Digs and Their Contribution to the History of Venetian Glass in the High Middle Ages," *Bruxelles Congrès,* 239.1–8; idem (supra n. 51) 68ff. G. R. Davidson, "A Medieval Glass-Factory at Corinth," *AJA* 44 (1940) 297ff.; F. R. Matson, "Technological Study of the Glass from the Corinth Factory," ibid., 325ff. (11th–12th C.).

121. For the shipping of raw materials cf. also Perrot (supra n. 77) 258.3.

comes from the level of the "Hellenistic Steps" (which are beneath BS E 14 to E 16) but somewhat east of them, under BS E 17. The piece seems to antedate the Byzantine period (*BASOR* 174, 47–50) and does not resemble any of the pieces previously discussed.

A few objects not unlike the Middle Byzantine cullet, **781–788**, came to light at levels that are clearly Early Byzantine.[122] As they were all found in or near Syn—the majority come from localities close to BS— one might be tempted to identify them as trade items in the Byzantine Shops. Glass melted into conveniently shaped ingots could have been easily transported to smaller workshops outside of Sardis having more primitive facilities.

Sections of flat cakes of bluish glass (**729** and **730**) —not necessarily disc-shaped—come from Syn Porch (*BASOR* 174, 47) which contained the remains of goblets, lamps (type 3), and cullet; these fragments are datable to the fourth to sixth century on the evidence of hundreds of coins found in the Syn Porch area (*BASOR* 170, 38ff.; 174, 47). The cakes might have been made at a factory that produced, among other fabrics, the aquamarine fabric 1 glass.

A section of another cake (**731**) of blue glass comes from Syn and most likely belongs to a period after A.D. 400; wasters were found close by (**735**; *BASOR* 174, 30ff.).

Finally, a section of a cake of fabric 1 glass (**732**) was found at Syn Fc where glass fragments of goblets and bowl lamps of type 2 also came to light. These sherds are Early Byzantine. The disruption characteristic, particularly within the southeastern part, of Syn Fc makes a more precise dating impossible.

Very few drippings—droplets fallen to the floor during manufacture—were found. One (**733**), is unstratified. A few others (**734**) belong to a group of wasters, including cullet, from the Byzantine period of BS W 13 (*BASOR* 157, 34).

Among the other finds there are a few irregularly shaped pieces that might be wasters or, less likely, sections of broken vessels badly misformed during a subsequent conflagration. All three (**735–737**) were found at Syn (*BASOR* 174, 30ff.; 177, 19). While the stratification of **735** and **736** is not precise enough to allow reliable dating, **737** belongs to the pavement underneath the fallen blocks which formed part of the fortification probably erected in the sixth century.

Pieces of Cullet (and/or Glass Fragments from Crucibles)

Most appear black (probably green or aquamarine); some are porous and coated with ceramic-like material.

712 BE-H 1964 E9-E12.5/N45-N53 *98.20 down.

713 BE-S 1964 E20-E26/N34-N40 *99.50.

714 *Pl. 17.* E of Syn Fc 1963 E115-E118/N3-N5 *98.00-96.75.

715 BE-C 1966 E31.9-E33.6/N13.09-N15.15. *97.50 down.

716 BE-B 1966 E26-E28/N7.50-N9.90 *97.40-96.40.

717 BS E 5, 1967 E35-E38/S0-S4 *98.23-97.50.

718 Lumps of cullet(?). Greenish-black; porous, pumice-like; irregular.
PN 1964 W237-W247/S347-S354 *88.60-88.00.

719 HoB 1968 W35-W42/S90-S110 to *100.30.

Glass or Glaze

Most of the pieces come from crucibles. They are, in part, of a ceramic material coated with glass or layers of glaze in varying thickness. Most are green, appearing black.

720 PN 1963 W227-W229/S360, S corner of room H, *89.45-89.30.

721 BS E 4, 1967 E29-E33/S1-S4 *98.05.

722 *Pls. 17, 28.* Section of crucible. Yellowish-green, bubbly glass attached to ceramic body that has a smooth, hard layer followed by a porous, disintegrating layer. The int., glass-coated surface slightly concave (curvature of int. of crucible).
Max. dim. 5.5; Th. glass layer 0.2, of hard ceramic body 0.3, of outer ceramic body ca. 0.5.
BE-B 1966 E27-E29/N5-N7 *96.80-96.40.

723 Many sections of crucible(?). Ext. slightly convex and vitrified (covered with glassy layer). Object may be the portion of a burned brick, as suggested by L. J. Majewski.
MTE 1964 E65-E70/S155-S160 *109.00-108.50.

724 Curved section of crucible(?); convex ext. glazed in brownish-green with lumpy surface. Int. appears to have layer of metal (bronze?).
Max. dim. 9; est. diam. ext. 10; Th. 2, of "bronze" layer 0.2.
MTE 1964 E68/S160 *109.20.

---

122. Similar cakes were found at Jerash: Baur-Kraeling, 546 (4th–5th C.). For similar cakes of later date cf. R. J. Charleston, "Glass 'Cakes' as Raw Material and Articles of Commerce," *JGS* 5 (1963) 54–67.

725   Many lumps (see *Pl. 17* for one example) of pumice-like, disintegrating ceramic material, perhaps from crucibles or kilns/furnaces (for firing ceramics or glass). They show vitrification and devitrified glass respectively. Glass appears black but is actually bluish-green (fabric 1?); it is often not homogeneous, having distinct sections in blue and in green.
Some lumps 10 by 8 by 8.
BE-C 1966 E24.50/N12.50 *96.80.

726   Flat layer of ceramic material (from crucible?) with irregularly shaped glass (cullet) attached to it. Blue.
3.7 by 1.3.
BS E 17, 1963 E101-E103.5/S0-S2.30 *93.85-93.30.

727   Ceramic element (from crucible?); curved; glazed in reddish and green.
MTE 1964 E60-E65/S148-S152 *112.00-110.00.

728   Similar to **727**; partly glazed; glass is light green.
Max. dim. 5 by 3.1, Th. of glassy layer up to 0.5.
HoB 1966 W22-W25/S124-S127 *100.20-99.90.

Sections of Glass Cakes

Flat; curved rims do not necessarily indicate that the cakes were disc-shaped. The thickness tapers to smooth rim.

729   *Pl. 17*. Blue changing to turquoise-blue; weathered scum.
Max. dim. 6.5, Th. 0.7.
Syn Porch 1962 E117-E119/N3.5-N4.5, in angle of walls, *97.50-97.00.

730   Similar to **729**.
Th. 1.
Syn Porch 1963 E118-E120/N2-N3 *98.00-97.50.

731   Blue.
Max. dim. 3.7, Th. 0.6.
Syn MH 1963 E75-E80/N16-N18 *97.50-96.75.

732   Lump of flat, irregularly shaped glass, aquamarine (fabric 1?).
Max. dim. 3.2, Th. 0.5-0.8.
Syn Fc 1963 E110-E115/N1.50-N6.00 *97.50-97.25.

Drippings and Wasters

733   *Pl. 17*. Tear-shaped dripping. Appears black (greenish?); decolorized, scum.
P.H. 2.3, diam. 1.1.
MTW 1967 W20-W35/S140-S145 to *102.00.

734   Group of wasters: drippings, cullet.
BS W 13, 1959 W55-W57/S1.70-S4.40 *98.00-97.00.

735   Waster? Fabric 1; decolorized, white scum. Irregular.
P.H. 4.
Syn MH 1963 E75-E80/N5-N18.5 *98.00-97.50.

736   Similar to **735**.
Max. dim. 2.3.
Syn 1963, S wall, E70-E75/N2-S1 *98.00.

737   Waster(?); irregular (might also be the remains of a melted object). Probably green (appears black); dirt, dull.
Max. dim. 3.
SE of Syn 1964 E125-E127/S10-S12 *97.00-96.50.

IV MIDDLE BYZANTINE GLASS

Remains of many glass bracelets were found in Middle Byzantine contexts while fragments of vessels datable to the same period are almost totally absent. All of the objects listed in this chapter belong to the late tenth to the thirteenth to fourteenth centuries. Unfortunately the stratigraphy is, in most cases, not clear, and that makes it almost impossible to distinguish between early and late examples within this period.

In addition to bracelets, the finds include window glass, cullet and cakes as well as a few fragments of Islamic vessels. The bracelets and perhaps the window panes were made locally while the vessels appear to be imports. The group of windows should be associated with Church E which is datable to the thirteenth century.[1]

BRACELETS

While the major portion of the Early Byzantine material comes from the general area of B and Syn, the bracelets listed here were found almost exclusively at PN and AcT; others come from CG and stray finds are from the upper levels at AT, L, HoB, B, and Syn.

The bracelets seem to have been made locally. This appears to be confirmed by the discovery of glass cakes (**781–788**) which could have served as the raw material for the bracelets. The workshops were certainly very small in comparison to the factory (or factories) of the fifth to early seventh centuries.

The diameters of the bracelets vary from ca. 6 to 9 cm. and most of them average about 8 cm. The thicknesses range from ca. 0.4 to 1.0 cm. They can be divided into two groups: decorated and undecorated bracelets. Within the series of decorated bracelets, four types can be distinguished.[2]

Decorated Bracelets

Type 1: Coil Twisted

The bracelet is twisted to form a spirally fluted coil. The basic color is usually dark green, rarely light milky yellow. The thin threads following the spiral

1. H. Buchwald, "Sardis Church E—A Preliminary Report," *Jahrbuch der österreichischen Byzantinistik* 26 (1977) 265–299.

2. Glass bracelets made in the ancient Near East are well represented in many collections; however, they are in most cases unpublished or their provenances are unknown. Exact parallels to those found in Sardis are not illustrated in archaeological reports, thus making it difficult to cite similar or identical pieces datable to the same time. For general remarks on glass bracelets and rings in pre-Roman and Roman times cf. Kisa, *Glas* I 138ff.; Vessberg, *Cyprus*, 213, mentions undecorated bracelets of the same period. For a survey of bracelets from Western sites cf. T. E. Haevernick, *Die Glasarmringe und Ringperlen der Mittel und Spätlatènezeit auf dem europäischen Festland* (Bonn 1960); idem, "Antike Glasarmringe und ihre Herstellung," *Glastechnische Berichte* 25 (1952) 212–215.

Sections of decorated bracelets (different from the types found in Sardis) in Lamm, *Samarra,* 106, 108 (9th C.); P. J. Riis, "Les verreries," in V. Poulsen et al., *Hama. Fouilles et Recherches 1931–1938. IV.2. Les Verreries et Poteries Médiévales* (Copenhagen 1957) 63, figs. 184–185 and 68, figs. 210–211 (10th. C. or slightly later). Close in time to the Sardis finds are the bracelets from Corinth datable to the 10th–12th C.: Davidson, *Corinth* XII 262–263, nos. 2138ff.

twist are either a combination of milky green and red, or single strands in yellow-green or red are used.[3]

738 G60.33:2758. Dark green with white-green and red threads.
Est. diam. 9, W. 0.5.
AcT 3–6/D–F, fill, *402.80-402.60, unstratified.
AJA 66, 12, no. 21, pl. 7:20.

739 Identical to **738**.
BE 1961 E18-E30/N85-N100, highway rubble, *101.00-100.00, unstratified.

740 Dark green with red.
Max. dim. 2.5.
AcT 1962 E4/N25 *404.80, unstratified.

741 Dark green or black with yellow-green.
Max. dim. 3.8.
PN 1963 W242-W248/S345-S349 *89.30-88.90. Found near glass cakes. Prior to mid-14th C., cf. *BASOR* 174, 14.

742 G60.5:2428. Milky yellow with red.
Est. diam. 8.
AcT trench A, 0–2/D–I, fill, *403.80-402.80. Found with **754**.

743 G58.1:9.
Trench S ca. W177.4-W179.4/S1281.5-S1283.5, unstratified, probably surface (cf. *Sardis* R1 [1975] 104–107).
Hanfmann, *JGS*, 54.

744 G58.3:225.
CG, near surface, unstratified.
Hanfmann, *JGS*, 54.

745 G58.54B:683.
L room B, surface to *99.00, fill. Found with Byzantine pottery.
Hanfmann, *JGS*, 54; *Sardis* R1 (1975) 111.

746 G58.64A:850.
L, unstratified.
Hanfmann, *JGS*, 54.

747 G58.65A:851.
L. Found with objects of Middle Byzantine occupation, including a 12th C. coin.
Hanfmann, *JGS*, 54.

748 G58.93A, B:1040. Twists in red-white and black(?)-red.
L room E *100.50-100.31, unstratified.
Hanfmann, *JGS*, 54.

749 Black(?) and white(?) coil.
CG 1969 W13.9/N28.3 *101.70.

750 *Pl. 18.* Creamy yellow, red and black coil.
CG 1969 E12.8/N59.35 *100.16.

Type 2: Bracelets with Two Exterior Threads

The bracelet is flat in cross section. On the exterior there is a double strip of contrasting color. The base color is either dark green or black, the "black" generally being very dark green. The decorative double strip consists of inlaid threads, apparently almost always in red and yellow-green.

751 *Pl. 18.* Dark green; 2 yellow-green stripes.
W. 0.5.
PN 1965 W285-W290/S320-S325 *88.70-88.20. Perhaps of the level of the early Islamic village, 14th C.

752 Dark (green?); a narrow red and a narrow yellow-green stripe.
Max. dim. 3.7, W. 0.7.
PN 1967 W265-W280/S320-S340 *88.60-88.20. Top level above late Roman level; Middle Byzantine intrusion(?), cf. *BASOR* 191, 9ff.

753 Appears black; grooved to form 2 distinct strands; on one strand a red, in the other a yellow-green stripe.
Max. dim. 5.3, est. diam. 8, W. 0.7.
BE-H 1964 E9-E12.7/N47-N53 *99.40 down; upper level, probably in rubble, cf. *BASOR* 177, 21ff., unstratified.

754 G60.6:2429. Black; central groove bordered by 2 red stripes.
Est. diam. 8, W. 0.9.
AcT trench A, 0–2/D–I, fill, *403.80-402.80. Found with **742**.

755 G60.45:2918. Black; one yellow-green and one wider red stripe.
Est. diam. 8.5, W. 1.0
AcT trench C, 1–4/B–E, fill below high wall, ca. *400.35, unstratified.

756 G58.64C:850. Blackish green with red stripe.
L, SW part of room A, unstratified.
Hanfmann, *JGS*, 53.

Type 3: Bracelets with Inlaid Rope

The bracelet is flat with an inlaid rope consisting of two spirally twisted strands of green and yellow or white and brown. The base color is black-amber (or very dark green?).

3. Coil bracelets of the 10th to mid-12th C. from Corinth: Davidson, *Corinth* XII, nos. 2143ff. A section of a coil bracelet with yellow thread, Riis (supra n. 2) 68, fig. 210.

757 *Pl. 18.* Black(?) with central green and yellow rope.
Max. dim. 3.8, W. 0.4.
PN 1967, exact findspot unknown.

758 Blackish amber with rope in white and brown.
Est. diam. 6, W. 0.5.
PN/E 1973 W203-W208/S357-S368, earth ramp, top soil,
*92.00-91.00, unstratified.

759 *Pl. 18.* G73.3:8239. Pale yellow inner band with applied white and yellow(?) coil on top.
PN/E W203-W208/S357-S368, top soil, *92.00-91.00.

Type 4: Bracelets with Inlaid Rectangles.

The bracelet is flat and generally has rectangular
patches of inlaid glass in contrasting colors. The base
color is black(?); the two most complete samples
found in this group (**760** and **761**) have inlays in a
combination of white, green, and yellow glass.

760 *Pl. 18.* Patch in white, green, and yellow.
Max. dim. 1.7, W. 0.7.
PN 1965 W275-W280/S325-S330 *88.20-87.70. Perhaps of
the level of the early Islamic village, 14th C., cf. *BASOR* 182,
25.

761 *Pl. 18.* Similar to **760**. Interconnected patches in
white, green, and yellow.
Max. dim. 3.5, W. 0.6.
PN 1967 W270-W280/S336-S340, S end, to *88.20. Top
level above late Roman level, Middle Byzantine intrusion(?),
cf. *BASOR* 191, 9ff.

762 Part of green patch.
Max. dim. 2.5, W. 0.8.
PN 1964 W252-W260/S345-S350, W of walls E and F,
*89.30-89.00. Found with plain rim of a fabric 3 vessel and
the section of a turquoise-blue bracelet. Early or Middle
Byzantine, cf. *BASOR* 177, 3–4.

Undecorated Bracelets

The bracelets are either flat or round in cross section. They are made of black, olive-green, and blue
glass. Their diameters vary from 7 to 8 cm.

Black Bracelet Sections

763 G61.31:4124. Complete bracelet (the only complete
one found at Sardis).
Diam. 7.3, W. ca. 0.6.
AcT W15-W17/N26, N of curving wall, ca. 0.80–1.00 below
ground level, unstratified.

764 HoB 1964 W1-W6/S120-S124 *99.85-99.40. According to the level this fragment could be datable to pre-Roman
times, cf. *BASOR* 177, 10, 13; however, it may also be a later
intrusion.

765 PN 1965 W275-W280/S325-S335 *88.70-88.50. For
dating cf. **751**.

766 Many bracelet sections. PN 1965 W275-W280/S325-
S330 *88.70-88.20. For dating cf. **751**.

767 G66.5:7037. Two bracelet sections.
Est. diam. 9, W. 0.8.
BS W 14, W67.10/S4.50 *97.95. Also found at this level was
a coin of "Anonymous Class A" (972–1028), *Sardis* M1 1971
no. 1145. For an apparently Early Byzantine bracelet section cf. **679**.
BASOR 186, 29.

Olive Green Bracelet Sections

768 *Pl. 18.* G62.7:4268.
Diam. 8, W. 0.5.
AcT E8.5/N20.5, at threshold unit 2, level Ib, *402.40, un-
stratified.

769 PN 1967 W274-W277.5/S330-S335 *88.35-88.10. For
dating cf. **761**.

770 Diam. 7.8.
AcT 1962 E8/N25 *402.90, unstratified.

771 PN 1965 W275-W280/S325-S330 *88.70-88.20. For
dating cf. **751**.

772 PN 1965 W285-W290/S320-S325 *88.50. For dating
cf. **751**.

773 G60.24:2729.
Est. diam. 7.5, W. 0.7.
CG W12-W18/S0-S5 *102.50-101.00. Industrial level, apparently late 10th-early 11th C. Found with coin of "Anonymous Class A" (972–1028), *Sardis* M1 (1971) no. 1157.

774 G58.41:553.
CG MAE, ca. *99.00.
Hanfmann, *JGS,* 53; *Sardis* R1 (1975) 136.

775 G58.54A:683.
L room B, surface to *99.00. Found with Byzantine pottery.
Hanfmann, *JGS,* 53.

Blue Bracelet Sections

776 Many sections.
CG 1959, MAW *99.00-98.20; MAE *99.50-98.00. Found
with tesserae, cf. *BASOR* 157, 40. 10th C. and/or later.

777 CG MAE 1958 *98.50. Possibly 10th C.
Hanfmann, *JGS*, 53; *Sardis* R1 (1975) 136.

778 G58.93C-E: 1040.
L room E *100.50-100.31. Found with twisted bracelets with
color strands, cf. **738–750**.
Hanfmann, *JGS*, 53; *Sardis* R1 (1975) 114.

779 *Pl. 18.* G73.2:8234. Bracelet section(?); heavy weathering. Blue; 2 rows of painted circles bounded by 2 lines and
interrupted by at least one pair of lines. (I have not seen the
object; the color of the "painted" decoration is unknown to
me.)
L. 3.2, W. 0.8.
BS, S of W 14, W40-W63/S7.30-S8.30 *98.00-96.50. May be
Middle Byzantine or possibly later.

WINDOWS

780 Group of ca. 30 window pane fragments. Pale blue
and black-blue, deep and blackish purple, pale yellow,
green, olive. Bubbly, generally well-preserved. a) One rim
fragment with a rounded edge; b) two fragments with large
and curved, folded edges; c) one fragment, with pontil
mark, the concave–convex central portion of a circular
pane ("bull's eye"). Some or all of the windows may have
been made in the crown glass technique.
Th. 0.15–0.6, 1 piece 1.0; est. diam. of 1 circular pane with
folded edge ca. 20.
PN grave 62.1.
PN/E 1972 W233-W236/S387-S390, Turkish houses within
SWB, *90.50 down.
PN/E 1972 W207-W210/S366-S369, 2nd apse trench.
PN/E 1973 W215-W216/S363-S364, W of face of MW 1,
*90.80 (bottom of trench). Found with gold leaf tessera.

CULLET

Raw material in the form of glass to facilitate the
melting process was also found in Middle Byzantine
context (cf. Early Byzantine cullet, **712–737**). During
this period, however, premelted glass in the shape of
cakes appears to have been the major, perhaps the
only, source for the making of objects in Sardis.[4] The
cakes, either prepared locally or, more likely, imported, were used by small workshops to fabricate
bracelets (cf. **738–779**). Some of the colors of the
bracelets and of the decorative threads and patches
inlaid in the bracelets are the same as those of the
cakes. In addition, cakes and bracelets were found,
more than once, in close proximity in levels datable to
the Middle Byzantine period. As no other glass objects were found save for a few stray vessel fragments

(cf. **789–796**), the cakes appear to have been used exclusively for making bracelets. The cakes consist of
flat pieces of glass in brick-red, green, and purple; the
latter two colors appear black in reflected light. The
thickness varies from 0.5 to 1.3 cm., with a median
thickness of 0.9 to 1.1 cm. Not a single complete specimen is recorded. In a few cases, however, the sections preserved show portions of curving rims that
serve as an indication that at least some of the cakes
were circular with diameters seemingly up to 11 cm.
(**784**). Other rim sections are straight (**781**) or show
an approximately right angle (**782**) which may mean
that cakes were also made in roughly rectangular
shape.

Most of the glass cake sections were found in the
"industrial" areas at PN; others come from B and CG.
781, from PN, is dated by coins of Bayazid ben Murad
(*terminus ante quem* of A.D. 1362) to a time before the
mid-fourteenth century (cf. *BASOR* 174, 14, 20ff.). A
series of about ten cake sections comes from an adjacent area at PN (**782–783**). They also seem to be datable by coins to the fourteenth century (*BASOR* 170,
15). No closely dated material came with **784**, a piece
that apparently was part of a circular cake (*BASOR*
174, 24). The brick-red of this section matches some
of the red thread decoration of the bracelets (cf., **752–
756**). **786** and **787** were excavated in the B area. The
former, apparently datable to a time after the collapse
of the vaults at BE-N (*BASOR* 187, 57–58), is very
likely mid-Byzantine; the date of the latter, on the
other hand, probably also a portion of a circular cake,
cannot be ascertained stratigraphically. The last section listed here, **788**, from CG, probably comes from
the industrial level that can be dated by coins to the
late tenth to early eleventh century (*BASOR* 162,
43ff.; *Sardis* M1 [1971] "Anonymous Class A," 128–
130).

The cake sections seem to date to the late tenth to
the first half of the fourteenth century. As the bracelets also belong to this period, we can be fairly certain

4. In Sardis glass cakes were more common in Middle, than in
Early Byzantine times (cf. **729–732**). Early (perhaps the earliest)
examples of glass cakes were found in Nimrud; they consist of seal-
ing wax red glass probably datable to the 2nd C. B.C.: A. von Saldern, "Glass from Nimrud," in M. E. L. Mallowan, *Nimrud and Its
Remains* (London 1966) II 633 and n.21 (the disc-like cake has a
diam. of ca. 16 cm. and is ca. 3.7 cm. Th.). For later glass cakes as a
commodity for glassmaking and enamel cf. particularly R. J.
Charleston, "Eighteenth Century Ingots of Glass," *Washington Congress*, 211–212; idem, "Glass 'Cakes' as Raw Material and Articles of
Commerce," *JGS* 5 (1963) 54–67. Cf. also P. N. Perrot, "The Excavation of Two Glass Factory Sites in Western Israel," *Bruxelles
Congrès*, 258.3.

that industrial activities at PN and other locations included the melting of (imported?) glass cakes to make simple bracelets.

Sections of Cakes and/or Cullet

Predominantly green glass appearing black.

781 Rim section; the rim appears to be straight.
Max. dim. 3.7, Th. 0.9–1.2.
PN 1963 W223-W229/S345-S355, N of SE wall of house with mosaics, *90.00-89.70.
BASOR 174, 14.

782 *Pl. 18*. Sections of cakes, including rim sections, and one corner rim section of angular cake. Surface irregular due to bubbles. Rim straight and curved.
Max. dim. 8, Th. 0.8–1.2.
PN 1962 W225/S370 *91.30.

783 *Pl. 18*. Three fragments from cakes, including one rim section. Brick-red with dark striations.
Th. 0.2–0.6.
PN 1962 W225/S370 *91.30.
BASOR 170, 15.

784 Cake may have been circular. Brick-red with dark striations (green?).
Max. dim. 5.7, est. diam. 11, Th. 0.5–0.8.
PN 1963 W236/S340 *89.20-88.95.

785 Rim section. Probably purple, appears black.
Max. dim. 5.5, est. diam. cake 8, Th. 0.8–1.3.
PN 1963 W236/S340, N of long Islamic wall, *89.20-88.45.

786 Rim section. Green, translucent, bubbly.
Max. dim. 4, est. diam. cake 8, Th. 1.0–1.3.
BE-N 1966 E16.80-E18/N92-N97 *97.50-96.80.

787 Section of circular cake. Green appearing black.
Est. diam. ca. 8.
B 1959, E side, ca. 2.00 E of wall, 15.00 N of ramp, ca. *98.00.

788 Green? Very thick.
CG 1960, corner HM and NS, floor *98.67.

VESSELS

The few vessels which belong to the period treated here seem to be the remains of decorated vessels that found their way to Sardis from regions most probably under Islamic rule.

The small neck fragment of a bottle with marvered and combed thread decoration flush with the surface (**789**) belongs to a well-known group of Islamic origin. Vessels of this type are usually made of deep purple or dark blue glass covered with white thread decoration combed to form spirals or feather patterns. They are generally dated in the tenth to twelfth century.[5] Unfortunately, the Sardis fragment is unstratified. A conical bottle neck with a heavy, applied spiral thread was certainly part of a globular bottle perhaps of the eleventh or twelfth century, parallels for which are usually found in the Eastern Mediterranean (**790**).[6] This piece is also unstratified. The unstratified fragment of a vessel, perhaps a bowl, with applied threads tooled to form a chain-like design **791**, could be possibly of late Roman Imperial date.[7] However, it seems more likely to be contemporary with the Middle Byzantine period in Sardis because it is related to Islamic vessels of the same date.[8]

The ring base of a large vase (**792**) of unknown type was found together with a coin of Michael VII (1071–1078). A handle with a rudimentary thumb rest (**793**) from the floor of the depressed area of Church E at PN may be an intrusion (*88.90) of the period when the church was turned into a workshop after the Islamic conquest (*BASOR* 174, 19). The next sherd (**794**) seems to have been part of a mold-blown Islamic vessel, probably a bottle, of a type popular in the Near East in the tenth to twelfth century.[9] An unstratified fragment of opaque red glass (**795**), the color of which resembles the red found in the Middle Byzantine bracelets (**738–779**), could be of this period or of later date. The last piece, which could have been the section of a vessel in the shape of an animal, is also unstratified (**796**); objects of this type are generally dated in the centuries around the turn of the first millennium.[10]

5. For various vessel shapes and decorative patterns in this category cf. Lamm, *Mittelalterl. Gläser,* pls. 29–32; Riis (supra n. 2) 61–69, figs. 179–209 (10th C. and later). Cf. also Saldern, *Slg. Hentrich,* nos. 339–342 (with additional refs.).

6. For bottles with conical neck decorated with an applied spiral thread cf. Lamm, *Mittelalterl. Gläser,* pls. 25–27; D. B. Harden, "The Glass Found at Soba," *Sudan Antiquities Service. Occasional Papers* 3 (1955) 69–70, pl. 20b (possibly 13th C.); *Smith Coll.,* no. 468 (ca. 10th–12th C.); Saldern et al., *Slg. Oppenländer,* no. 364 (ca. 11th–13th C.).

7. Barag, "Glass Vessels," pl. 34; *Smith Coll.,* no. 324.

8. Lamm, *Mittelalterl. Gläser,* pl. 25; A. von Saldern, "Ein gläserner Schlangenkorb in Hamburg," *Festschrift für Peter Wilhelm Meister* (Hamburg 1975) 56ff.; idem, *Slg. Hentrich,* nos. 327ff.

9. For bottles with honeycomb patterns cf., for example, Saldern, *Slg. Hentrich,* nos. 290ff. (with refs. to other pieces).

10. Lamm, *Mittelalterl. Gläser,* pls. 20ff.; Saldern *Slg. Hentrich,* no. 398 (with refs. to other animal vessels).

789 *Pl. 18.* G60.37:2762. Portion of concave neck and shoulder of bottle, spiral thread around neck, combed at shoulder. Pale amethyst, white sunken threads.
P.H. 3.
AcT trench E, 5–7/A–B, fill around wall, ca. *406.00, unstratified.
AJA 66, 12, no. 22, pl. 7:21.

790 *Pls. 18, 28.* G60.38:2763. Section of bottle neck tapering upwards; heavy applied spiral thread. Pale yellow-green; well-preserved.
P.H. 8.8.
AcT trench A, 12–13/A–B, ca. *404.00.
AJA 66, 12, no. 23, pl. 10:23.

791 *Pls. 18, 28.* G59.38a:1678. Lower portion of almost straight-sided vessel with ring foot. Applied and tooled thread decoration with a design of (apparently) interlocking chains. Pale yellow-green, bubbles; well-preserved.
Est. diam. at upper fracture ca. 6.
B ca. W16-W19/N16-N19, int. fill under arch W3, *99.50-98.00.
AJA 66, 11, no. 19, pl. 10:19.

792 G58.88:1026. Ring base from a large vase. Green (cf. Hanfmann, *JGS,* 53; not seen by the author).
Diam. 7.1.

L room B. Found with a coin of Michael VII (1071–1078), *Sardis* M1 (1971), no. 1181.
Hanfmann, *JGS,* 53.

793 Two-stranded handle (of bowl?) with rudimentary thumb rest. Green; well-preserved.
H. 3.8.
PN/E 1963 W215/S376, "pseudo-crypt," *88.90. Found with fragments of window panes of fabric 3.

794 *Pl. 18.* G60.48:2985. Small vessel fragment with curvature; pattern of mold-blown honeycomb decoration. Olive green tint; frosted.
Max. dim. 4, Th. 0.15.
HoB E5/S100 surface to *101.00.

795 G60.43:2855. Small vessel fragment. Opaque red-veined glass; well-preserved.
Max. dim. 1.8, Th. 0.1.
AcT trench E, 7–9/E–H, fill.

796 G60.39:2764. Lower section of vessel, perhaps an animal; 2 rudimentary feet(?) are preserved. Underside is pushed up slightly ("kick"). Purple.
P.L. 5.4, P.H. 1.9.
AcT trench E, 1–3/A–C, fill, *107.00-106.00.

V RING STONES AND BEADS

While the other material presented in this survey is grouped first by chronology and then by typology, ring stones and beads have to follow a different system. Practically all ring (or game?) stones catalogued here are indistinguishable from each other and many of them cannot be dated by means of stratigraphy. The beads are likewise catalogued in consecutive order instead of integrating them in the chapters on pre-Roman, Roman, Early, or Middle Byzantine glass. The problems of typology, chronology, and development of a multitude of types and their variants still await systematic investigation. Although many studies on beads have been published, it is still too early to make definite statements as to the dating and provenance of a given bead when found in a stratigraphically uncertain context. Also, undecorated beads of ordinary shape found in levels not safely dated can almost never be associated with a specific and well-known group since such beads were made for centuries without any change.

RING STONES AND ONE RING

A relatively large number of ring and/or game stones were found in Sardis (**797–819**). They generally have the shape of the segment of a sphere; two-fifths of a sphere seems to be the form most often used. The diameter ranges from 0.5 to 1.5 cm., with a median diameter of 1.1 to 1.3 cm. They were probably either set in rings of metal or glass, or they were used as game stones. The fabrics most frequently used are 1 and 3; stones of clear glass with a slight green tint, of amber, blue or black glass are rare.[1] The majority of the objects of this type seems to be of Early Byzantine date; some of those unstratified or associated with Roman (and even pre-Roman levels) may also conceivably be later.

One ring (**820**) is either Early Byzantine or Middle Byzantine (Islamic) as the level it comes from lies on top of the Hellenistic-Roman levels (cf. the ring **231** which is certainly Roman).[2]

Hemispherical or Flattened with Rounded Top

Mostly clear to greenish; dulled, and corroded. Diam. ca. 1.3 unless listed otherwise.

Roman or Earlier

797 G60.41:2840. Oblong, flat. Pale purple. 0.9 by 0.5.

1. Ring and game stones are particularly common in Roman times; they often have a device or a scene engraved on the flat underside, similar to gems. They were made in various colors, predominantly blue, greenish, clear, and yellowish. For general remarks on rings and ring stones cf. particularly Kisa, *Glas* I 141–142. Cf. Harden, *Karanis,* 284; Vessberg, *Cyprus,* 213, fig. 51:22 (Roman). Cf. also F. Fremersdorf, *Antikes, islamisches und mittelalterliches Glas . . . in den Vatikanischen Sammlungen Roms* (Vatican City 1975) nos. 328ff., 524ff., who identifies all of them as game stones.
2. For rings in general cf., for example, Kisa *Glas* I 138ff.; Harden, Karanis, 284; Crowfoot, *Samaria,* 420.

HoB E15/S100 *98.00. Post-Kimmerian, "not later than 500 B.C." (G. Swift).

798 G61.10:3297. Hemispherical. Greenish(?); decolorized.
Diam. 1.5.
HoB W10/S100-S105 to *99.95-99.85. Probably Hellenistic or Roman, fill over Lydian floor, cf. *BASOR* 166, 57.

799 Black scum.
Diam. 0.9.
PN 1965 W298-W301/S325-S330 *86.00-85.75.

800 Elongated; blue.
1.2 by 1.
HoB W20/S105 *100.60-100.20. Probably Roman.

801 HoB 1963 W38/S110-S115 *100.00. Roman?

802 HoB 1964 W12-W25/S117-S120 to *100.60-100.50. Probably Roman.

Early Byzantine

803 BS E 15, 1963 E90/S2.50 *93.00.

804 Tooled edge at base.
Diam. 1.2.
PN 1961 W280/S355 ca. *89.80. Probably Byzantine.

805 Silver iridescence.
Diam. 1.
Pa-W 1967 E33.80-E43.80/N95.70-N101.26 *96.83.

806 Blue.
Diam. 1.
Syn Fc 1967 E103.05-E106.55/N5.95-N9.05 *95.30-95.10. After A.D. 400.

807 Devitrified.
Diam. 1.5.
RT 1961 E7/S30 ca. *97.00. On top of Roman road.

808 G60.44:2858. Irregular, blackish.
N of B E33-E35/N125 ca. *96.00-95.00, N of rubble wall. Byzantine?

809 G61.1:3148. Black(?); decolorized, broken in two.
HoB E0/S95 *101.00. Probably Byzantine.

810 Ring stone or inlay. Rectangular, flat base. Green; frosted, iridescent.
L. 1.5, H. 0.5.
E of Syn 1966 E121.62/N1.95, on mosaic. After A.D. 400, on top of mosaic floor of 2nd half 4th C.

811 G67.3:7314. Elongated, almost oval in cross section, flat base. Clear; frosted. May not be glass.

L. 1.5, W. 1.
Syn Fc E106.95/N12.45 *96.07.

Additional Ring Stones

In type this group is similar or identical to the foregoing, but not safely datable through archaeological context. Most are of fabrics 1 and 3.

812 HoB 1962 W20/S90 *100.20-99.50, unstratified.

813 G67.1:7273. Aquamarine.
Diam. 1.1.
PN W268/S328.5 *88.50. Late Roman?

814 Diam. 1.2.
HoB 1961 W15/S95-S105 to *102.30, unstratified.

815 Diam. 1.2.
HoB 1968 W35-W42/S100-S110 to *100.20, apparently unstratified.

816 Diam. 1.4.
MTE 1964 E45-E50/S125-S130 *101.00-100.60, unstratified.

817 *Pl. 19*. G64.6:6434. Elongated; amber.
Diam. 1.4.
Syn MH E73-E75/N15 *96.40, on floor. After A.D. 400.

818 Diam. 1.1.
MTE 1964 E48-E51/S127-S129 *101.50-101.00, unstratified.

819 G64.4:6131. Glass? Greenish(?), smooth black surface.
Diam. 1.5.
MTE E64-E70/S150-S160 *109.00, unstratified.

Ring

820 Three-quarters of ring. Turquoise-blue; corroded.
Diam. 1.5, H. 0.7.
PN 1964 W250-W255/S340-S345 *89.70-89.40. Probably Byzantine, cf. *BASOR* 177, 3-4.

BEADS

The glass beads found in Sardis include decorated and undecorated examples, and date from the time between the late second millennium B.C. and the Early Byzantine period. Unfortunately there is as yet no general study of the history and typology of beads which would aid in the classification of the objects from Sardis. T. E. Haevernick has been working on the history of ancient glass beads for a long time and according to her such an all-inclusive survey would be practically impossible to compile. In a given period there may be as many as 20,000 beads with the same provenance but of numerous different types. To complicate matters even further, it appears that many types and their variants were not only made over an extended period of time, but were also frequently exported. We have therefore chosen not to cite parallel examples from other sites or from museum collections because such an enumeration would be of no help in placing the beads from Sardis in their proper contexts.

Decorated Beads

Eye beads are among the earliest glass objects from Sardis. Fortunately most of them are fairly well stratified. The earliest is a large, fragmentary, three-cornered bead (**821**) which comes from a level at HoB associated with the ninth to eighth century B.C. At this level were also found sherds of Greek Geometric pottery and their Lydian imitations (*BASOR* 186, 34). This dating is confirmed by T. E. Haevernick who puts this group in the eighth or seventh century. **822** represents the color scheme of a typical eye bead frequently found in Egypt and elsewhere in the Near East: yellow base color with blue and white eyes. It was discovered in a level of HoB that follows the Ionian invasion of 499 B.C. but antedates the Hellenistic-Roman level; it is, therefore, datable to the fifth or fourth century (*BASOR* 166, 8).

Two beads (**823**) may be of the same period or slightly later; the *terminus ante quem* is the destruction of this unit of PN by Antiochus III in 213 B.C. (*BASOR* 177, 6ff.). Another bead (**824**) from building C at HoB was found at a level with coins of Alexander the Great and Antiochus III (222–187) (*BASOR* 170, 10). One black bead with white dots (**825**) was part of the Hellenistic-Roman fill on top of the Lydian levels at HoB (*BASOR* 162, 12). Another, with discs and yellow wavy lines (**826**) is most probably Hellenistic.

Beads with wavy or straight thread patterns include divergent types. All examples of the first series have a decorative motif in common: inlaid threads arranged either in a wavy or zigzag pattern. Six or seven of them are of pre-Roman date. The first (**827**) with wavy lines in yellow on black, is definitely of the period of about 700 B.C. or slightly earlier as it comes from level III at PC. An almost identical bead (**828**) is unstratified. The same color combination is used in a biconical bead (**829**) found at HoB below the burnt floor that is associated with the Kimmerian raid of the

first half of the seventh century B.C. (*BASOR* 170, 6). Another small bead with wavy yellow lines, found nearby, seems to be of the same date (**830**). Typologically very close to **829** is a fragmentary bead (**831**) which was found just below the Hellenistic-Roman fill at HoB (*BASOR* 166, 5ff.). There is no doubt that it is Lydian although it is uncertain whether it is as old as the beads just mentioned.

A different type, **832**, is elongated and has three straight lines. According to the level, it appears to belong to the Persian period, that is the fifth or fourth century (*BASOR* 177, 4).

A third type is represented by an unstratified, biconical bead with a multicolored zigzag pattern (**833**); its date is uncertain as it comes from AcT.

Five beads with inlaid lines are datable to post-Roman times. The first, **834**, shows two wavy, intersecting lines and, coming from Syn Fc, appears to be Early Byzantine—late fourth century or later. The same period is represented by a tear-shaped bead with a white inlaid wreath (**835**); coins of Valens (A.D. 364–365), Honorius (393–423), Arcadius and Honorius (395–408), and Theodosius II (408–450) as well as Tiberius II (578–582) were found nearby in about the same level (*BASOR* 166, 16ff.).

A bead with three white lines, **836**, seems to be later in date as it was found in a level on top of the Byzantine graves at PN, thus dating it possibly in the Middle Byzantine period (*BASOR* 170, 16). A similar bead, **837**, lacks an excavation record.

Although found in a pre-Roman level (*BASOR* 191, 10ff.), **838** may be an intrusion of a much later date. White with blue ovals and a polished, shiny surface, it has every appearance of modern manufacture.

Two so-called melon beads with vertically incised lines came to light. According to the level, one (**839**) seems to belong to the Persian period, that is the late sixth to fourth century (*BASOR* 166, 19ff.). Although unstratified, **840** may be datable to the same time.

821 *Pl. 19*. G66.7:7072. One-third of three-cornered bead with white rings arranged concentrically around the corners. Black; dull.
Max. dim. 2.
HoB W1/S102 *96.00.

822 G61.13:3411. Yellow with 4 pairs of blue and white eyes; decolorized, frosted.
Diam. 1.2.
HoB W5/S95 *99.70.

823 G64.5:6388. Two slightly flat eye beads with 2 larger single, and 2 pairs of eyes. Probably blue, each eye in white with blue(?) pupil and another shade of blue(?) for center.

Heavily corroded and decolorized, fragile.
Diam. 1.3.
PN W300/S335, near unit 25, ca. *88.00-86.40.

824 *Pl. 19*. G62.14:4674. Three white eyes with blue centers. Slightly flat; blue, decolorized.
Diam. 1.
HoB building C W25/S85 *99.80.

825 G60.52:3034. Flattened; black, 3 white eye-dots; decolorized, corroded.
Diam. 1, H. 0.7.
HoB E23/S98 *100.20.

826 G61.17:3638. Discs and wavy lines in yellow; decolorized, black scum, fragmentary.
Diam. 1.5.
PN W250/S380 *89.00-88.70.
BASOR 166, 24.

827 *Pl. 19*. G60.4:2422. Spherical; dark green(?), appears black; wavy line framed by 2 plain lines in yellow.
Diam. 1.1.
PC zone 2, level III, *88.40. Found with fragments of pithoi with incised geometric pattern and "arrow" design (P60.152:2442; P60.153:2443).
AJA 66, 6, no. 1, pl. 5:1.

828 *Pl. 19*. G66.6:7064. Similar to **827**. Appears black.
Diam. 1.2.
HoB, from fill, ca. *97.50-96.00.

829 *Pl. 19*. G62.10:4402. Biconical, fragmentary. Appears black; in brilliant yellow: 2 wavy lines framed by straight lines, the central running along the greatest circumference.
Diam. 2.5.
HoB E5/S90 *97.50, lower than burnt floor.

830 *Pl. 19*. G62.12:4523. Spherical with pronounced edges at perforation. Dark green, wavy yellow line; corroded, mended.
Diam. 1.
HoB E5/S95, N edge, *96.30.

831 G61.29:3992. Biconical, fragmentary. Black or deep green, 2 yellow wavy lines flanking straight line in yellow; dull.
L. 1.7.
HoB E10/S90 *99.40.

832 *Pl. 19*. Elongated. Appears black, 3 white lines; shiny, polished surface.
L. 0.7.
PN 1964 W261-W265/S336-S338 *88.10-87.80.

833 *Pl. 19*. G61.16:3618. Slender, biconical; zigzag pattern in red, yellow (and blue?); decolorized.

P.L. 3.2.
AcT W22/N6.5 *402.00.

834 *Pl. 19*. G62.15:4699. Flattened. Probably green, 2 wavy lines in white crossing another; decolorized.
Diam. 1.5.
Syn Fc E98/N2 *96.80-96.70.

835 G61.2:3160. Tear-shaped. Aquamarine, inlaid wreath in white.
H. 1.5.
PN W265/S355, floor ca. *89.56.

836 *Pl. 19*. G62.1:4153. Slightly flattened. Appears black, 3 white lines, on one side pulled with pointed tool to form zigzag; dull.
Diam. 1.4, H. 1.
PN W235/S365 *91.40.

837 *Pl. 19*. One-half of flattened bead. Appears black, 3 white lines, on one side a vertical identation; decolorized.
Diam. 2, H. 1.4.
Chance find 1961.

838 *Pl. 19*. Flattened. White with diagonally arranged ovals in dark blue; shiny polished surface.
Diam. 0.6.
PN 1967 W255-W257/S340-S342 *87.60-87.00.

839 G61.30:3996. Spherical, vertically fluted like a melon. Light brown (amber); heavily corroded.
Diam. 1.
PN W245/S380, fill, *87.35-87.05.

840 *Pl. 19*. One-half of spherical bead, vertically fluted like a melon. Turquoise-blue; dull, frosted.
Diam. 2.
MTW 1964 W20-W25/S140-S145 to *102.00.

Undecorated Beads

A small number of undecorated beads of shapes other than the ordinary sphere or tube were found. A fragmentary, biconical bead (**841**) from HoB comes from the floor level that is apparently contemporary with the Ionian invasion of 499 B.C. (*BASOR* 166, 8ff.). A flattened, biconical bead (**842**), found a year later not far from the spot where **841** came to light, can be associated with Hellenistic coins of the late fourth to third century (Alexander, Antiochus III; *BASOR* 170, 10). Much later is a faceted bead (**843**); the level in which it was found contained many coins of the second and third century A.D. (*BASOR* 182, 16).

The last two beads of this series are late Roman or Early Byzantine. One shows a band of notches (**844**);

stratigraphic evidence indicates that it may be of the fourth or fifth century (*BASOR* 162, 40ff.). The other (**845**) is of biconical shape and, therefore, similar to those of pre-Roman times (cf. **841** and **842**); it was found in a pit underneath the floor of BS E 14 and thus must be earlier than the finds from this shop which are datable to the fifth and sixth century.

About two dozen beads of spherical or oval shape as well as a group of tubular beads were found in Sardis. According to stratigraphic evidence these very ordinary objects appear to have been made without any visible change throughout the first millennia B.C. and A.D. In the following, a sample list of more or less securely dated beads is given.

A spherical bead of yellow tinted glass (**846**) is probably the earliest piece of glass found in Sardis. It was discovered above the floor of the "Lower Burning Level" at HoB and may, therefore, be datable to the late second millennium B.C. (*BASOR* 162, 16). An apparently greenish spherical bead from HoB (**847**) must be dated to pre-Kimmerian times, namely the late eighth or early seventh century (*BASOR* 182, 9). Slightly later is a flattened bead of bottle green glass (**848**; *BASOR* 186, 32).

According to the excavator, A. Ramage, the section of an aquamarine colored bead (**849**) should be datable to the sixth century B.C. The stratigraphy of a ring-like bead (**850**) seems to place it in the late seventh or sixth century B.C. (*BASOR* 182, 11ff.). Slightly later, namely of the late sixth or fifth century, is a flat yellowish bead from PN (**851**; *BASOR* 191, 10ff.). A similar one, though spherical (**852**), is perhaps datable to a time prior to the Ionian destruction (*BASOR* 166, 8), while another (**853**) seems to postdate this invasion of 499 B.C. but is earlier than the Hellenistic period (*BASOR* 166, 8). The stratigraphy of **854** is less clear although it seems to be Persian or Hellenistic (*BASOR* 182, 11ff.). An elongated bead (**855**) of turquoise blue glass, found at HoB, belongs to the same period as **853**.

Less well dated is a greenish spherical bead (**856**) which, according to the late G. F. Swift III, can be associated with all types of pottery but is most probably not later than the third century B.C. Another spherical bead (**857**) was part of the Hellenistic-Roman fill at HoB (*BASOR* 162, 12).

Seven tubular beads (**858**) were found in a Hellenistic grave of the third or second century at Hacı Oğlan (*BASOR* 166, 30). Undoubtedly of Roman date is another tubular bead (**859**); coins of the second and third century A.D. were found in the same level. A group of additional undecorated beads of pre-Roman and Roman dates is listed under **860–866**.

Early Byzantine beads, in shape and size similar to the earlier specimens, are rare in Sardis. Those that were recorded are listed under **867–874**.

Pre-Roman and Roman

841 One-half of biconical bead. Greenish; white scum.
Diam. 1.
HoB 1961 W10/S91 *99.20, floor level.

842 G62.13:4554. Flattened, biconical. Turquoise-blue; corroded.
Diam. 1.
HoB W20/S90 *99.50-99.30.

843 Cut into diamond facets. Blue.
Diam. 0.8.
HoB 1965 W36-W39/S113-S120 *101.15-101.00.

844 G60.47:2948. Flattened, tooled to form vertical notches. Turquoise-blue; white scum.
Diam. 1.5.
LNH 1, E33-E35/N115-N125 *97.00-95.00.

845 *Pl. 19*. G63.1:4976. Biconical. Greenish; dark scum, one end fractured.
P.L. 4.8, est. L. 5.3, diam. 2.7.
BS E 14, E83-E90/S0-S3 *96.50-94.00.

846 G60.50:3002. Spherical. Yellow tint; decolorized.
Diam. 1.2.
HoB E10-E15/S105 over *95.41 floor.

847 *Pl. 19*. G65.3:6844. Approximately spherical. Greenish(?); silver iridescence.
Diam. 1.2.
HoB W0-W2.5/S99.50-S101.50 *97.50.

848 *Pl. 19*. G66.2:6995. Flattened. Bottle green; shiny surface.
Diam. 1.9, H. 1.2.
HoB W5-W9/S106-S109 *97.40-97.00.

849 One-half of flattened bead. Aquamarine; heavily corroded.
Diam. 1.3.
PN 1968 W266.40/S322.70, below lower water channel, ca. *86.00.

850 G65.2:6815. Ring-shaped. White scum.
Diam. 0.7.
HoB W12-W15/S110-S115 *97.90.

851 Flattened. Yellow tint; corroded.
Diam. 0.8.
PN 1967 W266.8-W267.3/S322.9-S323.9 *86.40-85.80.

852 G61.28:3926. Spherical. Yellowish.
Diam. 0.7.
HoB W11/S93 *98.90-98.85.

853 G61.12:3408. Spherical. Greenish(?); decolorized.
Diam. 0.6.
HoB W10/S95 *99.80.

854 Spherical. Black; partly exposed to fire.
Diam. 1.3.
HoB 1965 W15-W20/S110-S115 *99.00-98.80.

855 Elongated bead-like object. Turquoise-blue.
L. 0.7.
HoB 1961 E0/S90-S95 *99.90-99.80.

856 Almost spherical. Greenish; dull.
Diam. 1.3.
HoB 1968 W22-W26/S120-S124 *100.50-100.30.

857 G60.20:2725. Spherical. Brilliant iridescence.
Diam. 0.8.
HoB E10/S100-S105 *100.00-99.50.

858 G61.3:3161. Seven tubular beads. Decolorized, they now appear white, greenish, or black.
Average L. 0.6.
Hacı Oğlan, tomb 61.3, inside pelvic bone, 0.13 from S inside, 0.44 from upper edge, 0.33 from E side.

859 Tubular. Black; fractured.
L. 1.3.
HoB 1965 W25-W35/S120-S130 *101.90-101.60.

860 G68.1:7613. Ring-shaped. Black.
Diam. 1.1.
HoB W7-W8/S90-S92 *96.10.

861 Slightly flattened. Two colors, greenish.
Diam. 1.5.
MTE 1964 ca. E65-E75/S150-S165 *114.40.

862 G60.42:2850. Irregular. Turquoise-blue.
Diam. 1.
AcT trench B, 11–13/A–B, fill, *100.30.

863 Spherical. Black; dull.
Diam. 1.
HoB 1968 W35-W42/S90-S110, up to 0.40 below surface at ca. *101.50.

864 One-half of spherical bead. Turquoise-blue.
Diam. 1.2.
AhT 1967 E38/N0-N3, surface near lake.

865 Part of bead or inlay. Hemisphere in "Egyptian blue" color; porous, disintegrating.
Diam. 2.2, H. 1.5.
MTE 1964 E64-E71/S155-S160 *109.80-109.40.

866 G64.3:6130. Spherical. Black with 2 white dots.
Diam. 1.
MTE E69-E70/S150-S160 *109.00.

Early Byzantine

867 G60.23:2728. Spherical. Turquoise-blue; weathered.
Diam. 0.9.
CG W22-W26/N2-S5 *102.50.

868 Oval.
L. 0.5.
RT 1962 E28/S29 *96.75.

869 Spherical.
Diam. 0.4.
Syn Fc 1965, marble pile.

870 *Pl. 19.* Irregularly shaped, spherical object, melted bead(?). Black; cf. **871**.
Diam. 0.8.
BS E 2, 1967 E12-E13/S2-S2.50 *96.50-96.40.

871 *Pl. 19.* Same as **870**. These pieces do not seem to be drippings from glass manufacture but deformed objects, possibly beads.
Diam. 1.
BS E 13, 1962 E70-E73/S4 *96.70-96.20.

872 Two fragments of possibly spherical bead. Turquoise-blue.
RT 1961 E0-E5/S7-S10, pavement surface, *97.60.

873 Tube-shaped.
L. 1.2.
SE of Syn 1964 E125-E126/S10-S12 *97.25-96.75.

874 G63.5:5190. Tear-shaped. Blue; silver iridescence.
L. 1.3.
Syn Fc E114-E115/N4-N6, fill above floor, *97.50-96.75.

CONCORDANCE

Inventory Number	Catalogue Number	Plate Number
Pieces from the excavations		
G58.1:9	743	
G58.3:225	744	
G58.14:420	135	22
G58.15:466	560	15
G58.17A, B:469	681	
G58.18:470	681	
G58.22:408	402	
G58.23:409	403	
G58.27:387b	404	
G58.34:528	310	
G58.38:538	260	
G58.39:547	182	
G58.41:553	774	
G58.44:586	311	
G58.45:587	311	
G58.46:611	311	
G58.47:612	654	
G58.54A:683	775	
G58.54B:683	745	
G58.60:796	563	
G58.64A:850	746	
G58.64C:850	756	
G58.65A:851	747	
G58.68:865	608	
G58.71:866	438	
G58.74:879	655	
G58.77:922	439	
G58.83:1009	607	
G58.84:1010	6	
G58.88:1026	792	
G58.93A, B:1040	748	
G58.93C-E:1040	778	

Inventory Number	Catalogue Number	Plate Number
G58.95:1048	593	
G59.5:1189	644	
G59.16:1262	581	
G59.21a:1328	578	
G59.22:1340	445	13, 24
G59.25:1341a	569	15
G59.28:1374	568	
G59.29B:1382	312	
G59.30:1390	279	
G59.32a:1636	247	11
G59.33a(γ):1640	60	2
G59.35a:1658	71	20
G59.37:1456	319	12
G59.38a:1678	791	18, 28
G59.39:1472	125	21
G59.40:1480	351	24
G59.42:1483	472	13, 25
G59.42a:1769	320	
G59.44a:1810	70	3, 20
G59.45a:1843	119	6
G59.46:1530	34	2
G59.46A:1850	134	7
G59.46B:1850	130	7
G59.47a:1851	118	5
G59.48a:1855	131	7, 22
G59.49a:1885	642	15
G59.53:1620	80	
G59.57:2001	350	
G59.58:1804a	374	12, 24
G59.61:2122	62	3, 20
G59.62:2150	497	
G59.66:2189	287	23
G59.67:2193	313	

Inventory Number	Catalogue Number	Plate Number	Inventory Number	Catalogue Number	Plate Number
G59.68:2194	367	12	G61.23:3775	518	27
G59.71:2217	510		G61.25:3804	2	1
G59.72:2218	647	15, 28	G61.26:3882	3	1
G60.1:2242	136	7, 22	G61.27:3910	103	
G60.3:2393	26	2, 20	G61.28:3926	852	
G60.4:2422	827	19	G61.29:3992	831	
G60.5:2428	742		G61.30:3996	839	
G60.6:2429	754		G61.31:4124	763	
G60.7:2633	269	11	G62.1:4153	836	19
G60.9:2714	8	20	G62.2:4160	564	15, 27
G60.10:2715	13		G62.4:4162	669	28
G60.11:2616	14	1, 20	G62.5:4163	668	16, 28
G60.12:2717	29		G62.6:4164	670	
G60.14:2719	96	4	G62.7:4268	768	18
G60.19:2724	222		G62.8:4227	240	11
G60.20:2725	857		G62.9:4340	671	16
G60.23:2728	867		G62.10:4402	829	19
G60.24:2729	773		G62.11:4487	519	14, 27
G60.25:2730	128	6	G62.12:4523	830	19
G60.26:2731	194		G62.13:4554	842	
G60.27:2732	127	6	G62.14:4674	824	19
G60.31:2736	123	6, 21	G62.15:4699	834	19
G60.32:2742	129	7, 21	G62.16:4744	77	21
G60.33:2758	738		G63.1:4976	845	19
G60.36:2761	104	4, 21	G63.2A:5034	248	
G60.37:2762	789	18	G63.2B:5034	249	11
G60.38:2763	790	18, 28	G63.3B:5114	214	
G60.39:2764	796		G63.4:5130	200	10
G60.41:2840	797		G63.5:5190	874	
G60.42:2850	862		G63.6:5234	132	7, 22
G60.43:2855	795		G63.7:5249	667	16, 28
G60.44:2858	808		G63.8:5253	188	22
G60.45:2918	755		G63.9:5255	230	
G60.47:2948	844		G63.11:5302	662	16
G60.48:2985	794	18	G63.13:5309	645	
G60.50:3002	846		G63.14:5373	473	13
G60.51:3015	76		G63.15:5376	126	6
G60.52:3034	825		G63.16:5407	124	6
G61.1:3148	809		G63.18:5785	63	3, 20
G61.2:3160	835		G63.19:5829	444	24
G61.3:3161	858		G63.20:5880	464	
G61.7:3181	232		G63.21:5903	395	13, 24
G61.8:3196	133	7	G64.1:6044	86	
G61.9:3268	422		G64.2:6129	215	10, 22
G61.10:3297	798		G64.3:6130	866	
G61.12:3408	853		G64.4:6131	819	
G61.13:3411	822		G64.4a:6157	97	4
G61.14:3412	696		G64.5:6388	823	
G61.16:3618	833	19	G64.6:6434	817	19
G61.17:3638	826		G64.7:6519	397	24
G61.18:3770	237	11, 23	G64.8:6520	338	
G61.19:3771	481	14	G64.9:6558	624	27
G61.20:3772	490		G65.1:6786	451	25
G61.21:3773	646		G65.2:6815	850	
G61.22:3774	487	26	G65.3:6844	847	19

Inventory Number	Catalogue Number	Plate Number
G66.1:6975	562	15
G66.2:6995	848	19
G66.3:7012	570	15
G66.4:7019	658	
G66.5:7037	767	
G66.6:7064	828	19
G66.7:7072	821	19
G66.8:7088	1	1
G66.9:7140	5	1
G66.10:7261	565	15
G66.11:7263	234	11, 23
G67.1:7273	813	
G67.2:7287	672	28
G67.3:7314	811	
G67.4:7315	673	28
G67.5:7401	674	16
G67.6:7444	571	15
G67.7:7470	116	5, 21
G67.8:7488	111	
G67.9:7489	114	5, 21
G67.10:7490	110	5, 21
G67.11:7520	112	5, 21
G68.1:7613	860	
G68.2:7686	235	23
G68.3:7736	488	14, 26
G72.1:8181	509	14, 26
G73.1A:8233	400	24
G73.1B, C:8233	236	
G73.1D, E:8233	246	11, 23
G73.1F:8233	246	
G73.2:8234	779	18
G73.3:8239	759	18
G73.7:8264	314	12
G73.8:8267	233	10
J64.1:6108	231	
Seal58.1:986	218	22

Inventory Number	Catalogue Number	Plate Number
Non-excavated pieces from the area of Sardis		
NoEx68.5	216	10
NoEx68.17.1	152	
NoEx68.17.2	148	
NoEx68.17.3	156	8
NoEx68.17.4	158	
NoEx68.17.5	159	8
NoEx68.17.6	145	
NoEx68.17.7	170	9
NoEx68.17.8	163	8
NoEx68.17.9	167	9
NoEx68.17.10	161	
NoEx68.17.11	171	
NoEx68.17.12	168	9
NoEx68.17.13	143	
NoEx68.17.14	176	
NoEx68.17.15	165	8
NoEx68.17.16	151	
NoEx68.17.17	144	
NoEx68.17.18	155	8
NoEx68.17.19	146	
NoEx68.17.21	162	
NoEx68.17.22	175	9
NoEx68.17.23	172	9
NoEx68.17.24	147	
NoEx68.17.25	164	
NoEx68.17.26	150	
NoEx68.17.27	154	
NoEx68.17.28	173	9
NoEx68.17.29	160	8
NoEx68.17.30	149	
NoEx68.17.31	142	8
NoEx68.17.32	174	9
NoEx68.17.33	157	
NoEx68.17.34	153	
NoEx68.17.35	166	8
NoEx68.17.36	169	9
NoEx71.19	79	3

INDEX

References to catalogue entries are in bold type.

ILLUSTRATIONS

PHOTOGRAPHIC CREDITS

Illustrations of excavated pieces and of plans are the property of the Archaeological Exploration of Sardis, Fogg Art Museum, Harvard University. Unless listed below, illustrations of comparative material are from museum photographs or from photographs kindly furnished by the owner of the object. We are grateful to those museums and collectors through whose courtesy photographs of objects in their collections are published.

Pls. 20, 21   Reproduced from C. Isings, *Roman Glass from Dated Finds* (Groningen and Djakarta: J. B. Wolters 1957) forms 16, 29 and 42 by permission of Wolters-Noordhoff bv, Groningen.

Pl. 22   Reproduced from S. Loeschcke and H. Willers, *Beschreibung römischer Altertümer, gesammelt von Carl Anton Niessen,* 3rd ed. (Cologne: Greven and Bechtold 1911) pl. 23, no. 274 by permission of Greven Verlag, Cologne.

Pl. 22   Reproduced from Donald B. Harden, *Roman Glass from Karanis Found by the University of Michigan Archaeological Expedition, 1924–1929* (Ann Arbor: University of Michigan Press 1936) pl. 14, nos. 221 and 304 by permission of the Kelsey Museum of Ancient and Mediaeval Archaeology, University of Michigan.

Pls. 23, 25   Reproduced from D. Barag, "The Glass" in M. W. Prausnitz, *Excavations at Shavei Zion: The Early Christian Church* (Rome: Centro per le An-tichità e la Storia dell'Arte del Vicino Oriente 1967) fig. 16 by permission of the author and the Centro per le Antichità e la Storia dell'Arte del Vicino Oriente, Rome.

Pl. 23   Reproduced from P. V. C. Baur, "Glassware" in Carl H. Kraeling ed., *Gerasa: City of the Decapolis* (New Haven: American Schools of Oriental Research 1938) fig. 20 and fig. 22 no. 380 by permission of the American Schools of Oriental Research.

Pl. 23   Reproduced from Grace M. Crowfoot and D. B. Harden, "Early Byzantine and Later Glass Lamps," *JEA* 17 (1931) pl. 29 by courtesy of the Egypt Exploration Society.

Pl. 25   Reproduced from D. Barag, "Glass Vessels of the Roman and Byzantine Periods in Palestine" (Diss. Jerusalem 1970) pls. 11, 43 and 45 by permission of the author.

Photographs and drawings of excavated vessels are approximately half actual size, with the exception of some small fragments. Seals and gems are reproduced approximately one to one or one to two.

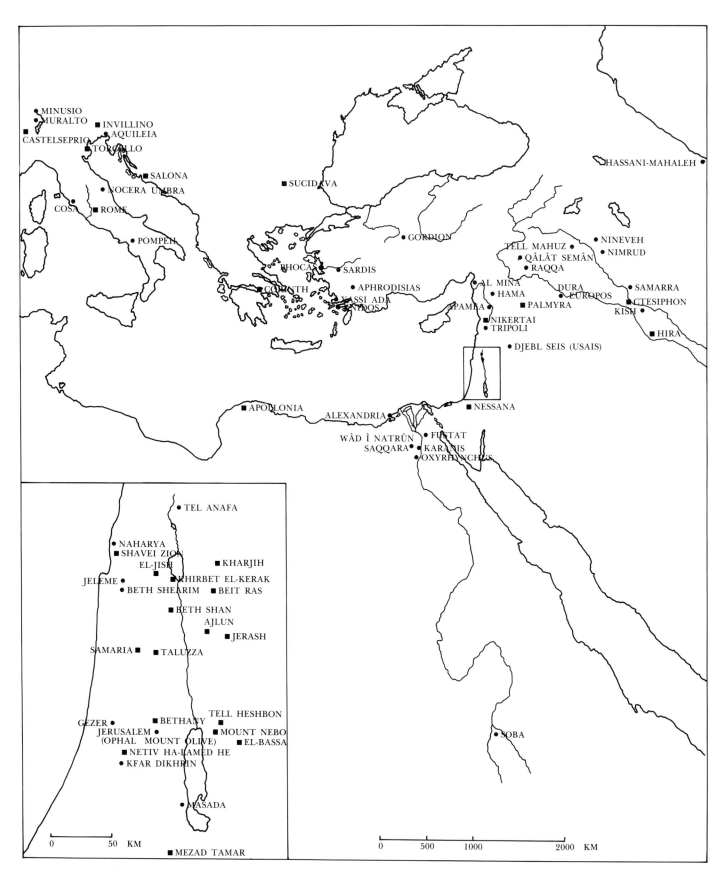

Plan I. Map of Major Glass Sites in Europe and Asia Minor. Those marked with a ■ have yielded primarily Early Byzantine finds.

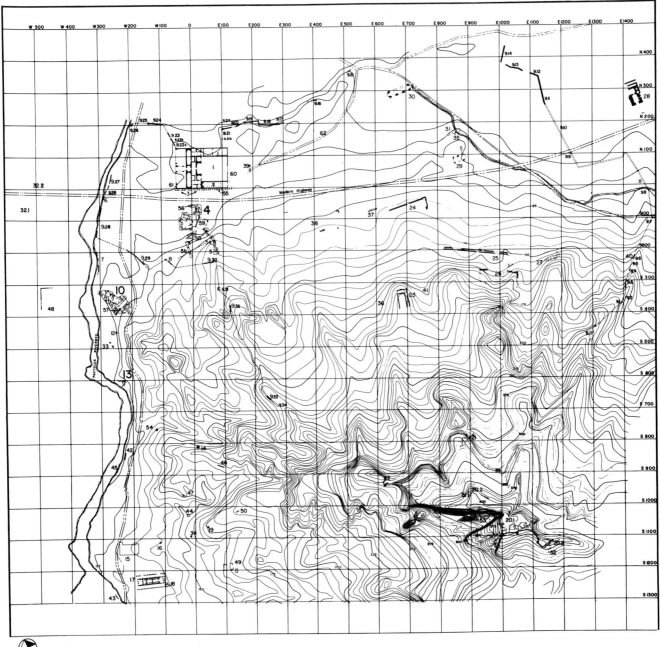

 mag. N

0 50 100 150
meters

1. Gymnasium-Bath
2. Synagogue
3. Byzantine Shops
4. House of Bronzes
5. Upper and Middle Terraces (a, b)
6. Roman Bridge
7. Pactolus Industrial Area
8. Southwest Gate
9.1-9.34. Byzantine City Wall
10. Pactolus North
11. Churches E and EA
12. Peacock Tomb
13. Pactolus Cliff
14. Pyramid Tomb

15. Expedition Headquarters
16. Northeast Wadi
17. Temple of Artemis
18. Church M
19. Kâgirlik Tepe
20.1 Acropolis Top
20.2 Acropolis North
20.3 Acropolis South
21. Acropolis Tunnels
22. Flying Towers
23. Byzantine Fortress
24. Building A
25. Stadium
26. Theater
27. Hillside Chambers
28. Bath CG
29. Building D (Byzantine Church)

30. Building C (Roman Basilica)
31. Mill
32.1 Claudia Antonia Sabina Tomb
32.2 Painted Tomb
33. Brick Vaulted Tombs
34. Roman Chamber Tomb
35. Road under Mill
36. Road to Byzantine Fortress
37. Vaulted Substructure
38. Roman Agora
39. Rubble Walls East of Gymnasium
40. Odeum Area
41. Foundations
42. Hypocaust Building
43. Marble Foundation
44. Minor Roman Building
45. Rubble Wall

46. Wall
47. Brick Vaulted Tomb
48. Walls
49. Butler's House
50. Shear's Stoa
51. Lydian Walls (AcN)
52. Pre-Hellenistic Walls (AcS)
53. Holes in Acropolis Scarp
54. Şeytan Dere Cemetery
55. Hellenistic Steps
56. Hellenistic Tombs
57. Street of Pipes
58. HoB Colonnaded Street
59. Building R and Tetrapylon
60. East Road
61. West Road?
62. Conjectured Ancient Road

Plan II. Site plan with excavations and ruins of Sardis.

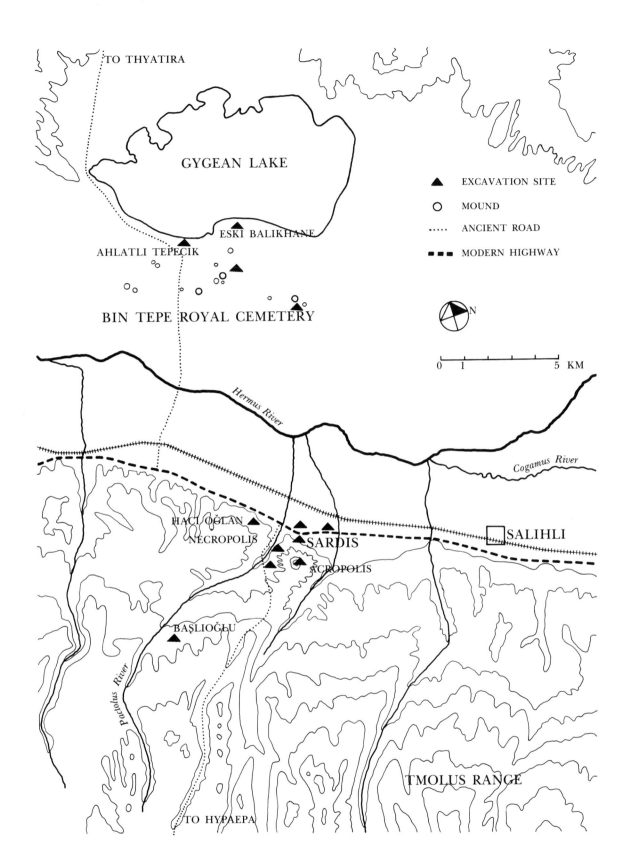

TO THYATIRA

GYGEAN LAKE

▲ EXCAVATION SITE

◯ MOUND

..... ANCIENT ROAD

▰▰▰ MODERN HIGHWAY

ESKI BALIKHANE

AHLATLI TEPECIK

N

BIN TEPE ROYAL CEMETERY

0   1   5 KM

Hermus River

Cogamus River

HACI OĞLAN
NECROPOLIS

SARDIS

SALIHLI

ACROPOLIS

BAŞLIOĞLU

Pactolus River

TMOLUS RANGE

TO HYPAEPA

Plan III. Map of the Sardis Region.

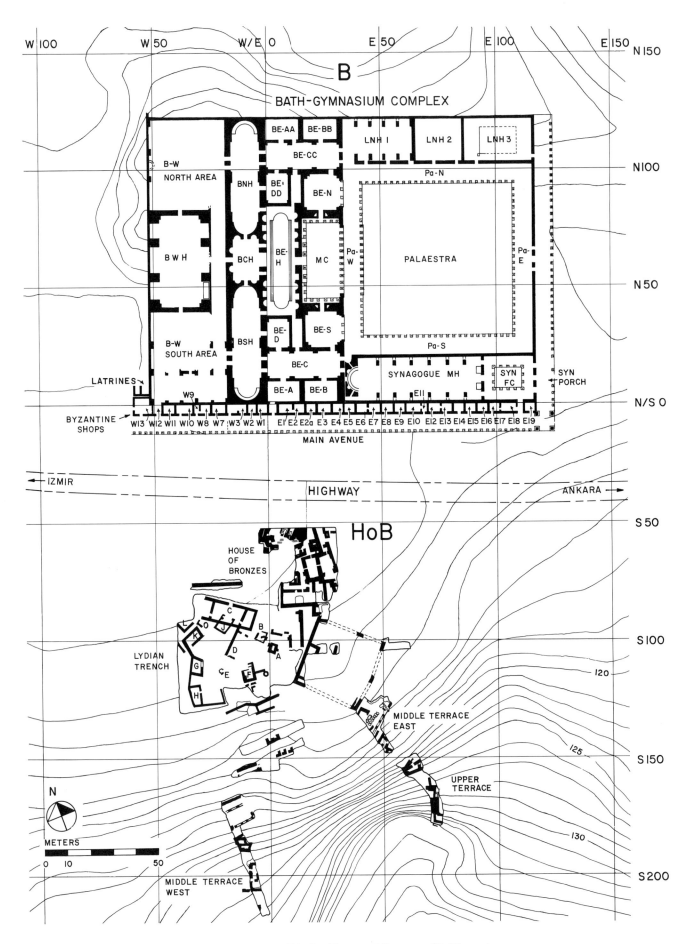

Plan IV. Plan of the Bath-Gymnasium complex (B), the House of Bronzes (HoB), and terrace trenches (UT, MTE, MTW).

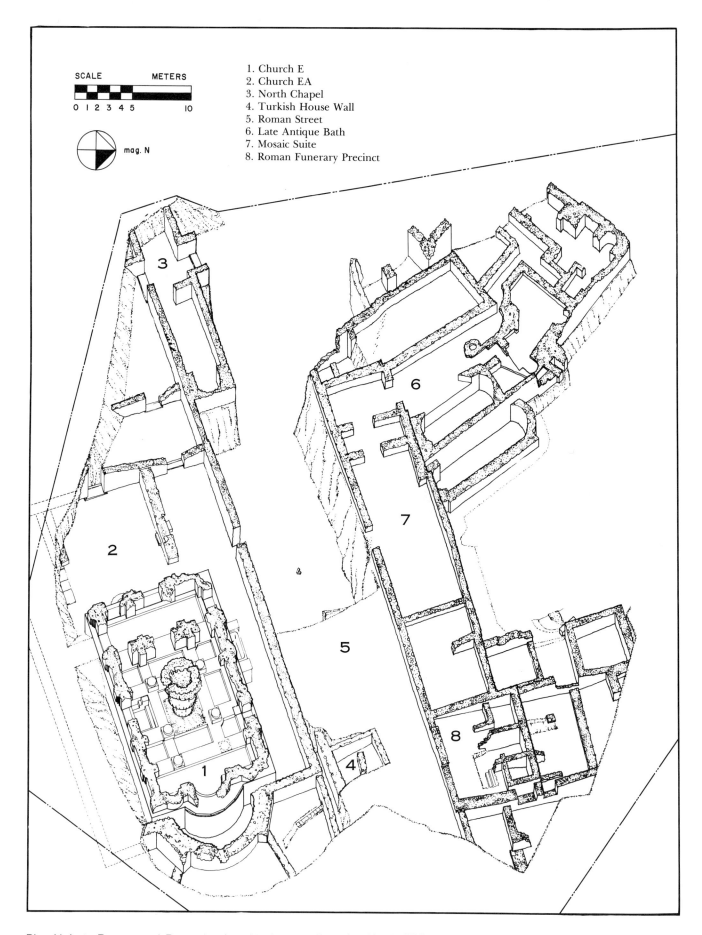

SCALE   METERS

0 1 2 3 4 5   10

mag. N

1. Church E
2. Church EA
3. North Chapel
4. Turkish House Wall
5. Roman Street
6. Late Antique Bath
7. Mosaic Suite
8. Roman Funerary Precinct

Plan V. Late Roman and Byzantine Levels of sector Pactolus North (PN).

PLATE 1   Pre-Roman and Roman

2

3

1

4

Core alabastron, Eastern Mediterranean,
6th–4th C. B.C. Oppenländer Collection,
Waiblingen (2121).

22

14

Hellenistic bowl, "Alexandria (?)", 2nd–early 1st
C. B.C. Museum für Kunst und Gewerbe, Hamburg (1975.63g).

25

Patella cup, possibly Syrian-Palestinian Coast, late 1st C. B.C.–1st
A.D. The Corning Museum of Glass (74.1.20).

26

Skyphos, Near East (Alexandria?) or Italy, late Hellenistic or Early Imperial. Museum of Fine Arts, Boston (50.2285, Bequest of Charles B. Hoyt).

34

Pillar-molded bowl, probably Eastern Mediterranean, 1st C. A.D. Kunstmuseum, Düsseldorf (P1966–76).

43

Zarte Rippenschalen, Eastern Mediterranean or Italy, A.D. 60–100. Oppenländer Collection, Waiblingen (22603, 22613, 2687).

53

56

57

60

Bowl and beaker with grooves, Eastern Mediterranean, late 1st–early 2nd C. A.D. Oppenländer Collection, Waiblingen (2698 and 2315).

PLATE 3   Roman

62

63

67

64

Bowl with facet decoration, Eastern Mediterranean, 3rd–4th C. A.D. Oppenländer Collection, Waiblingen (2658).

69

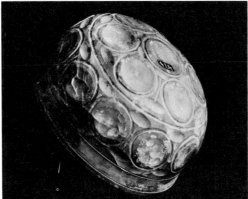

70

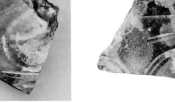
72

73

Hemispherical bowl with cut decoration, Cologne, ca. A.D. 200. Römisch-Germanisches Museum, Cologne (474).

79

80

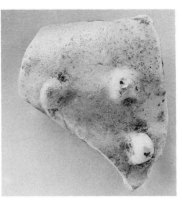
81

Bowl with warts, Rhineland, 3rd C. A.D. Rheinisches Landesmuseum, Bonn (6662).

91

92

Jug with "nip't diamond waies,"
probably Syria-Palestine, 3rd C.
A.D., Kunstmuseum, Düsseldorf
(P1972–123).

96

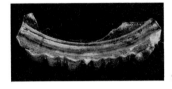

97

Bowl with crimped bands, Eastern Mediterranean, 2nd–
3rd C. A.D., Oppenländer Collection, Waiblingen (2630).

104

105

107

108

PLATE 5   Roman

110

112

113

114

115

116

117

118

124

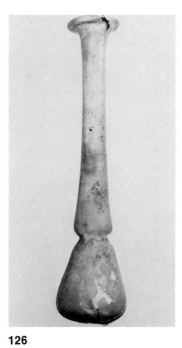

119

123

126

127

128

PLATE 7   Roman

129

130

131

132

133

134

136

142

155

156

159

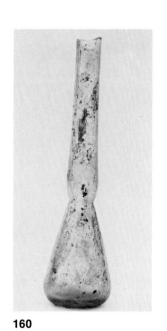

160

163

165

166

PLATE 9   Roman

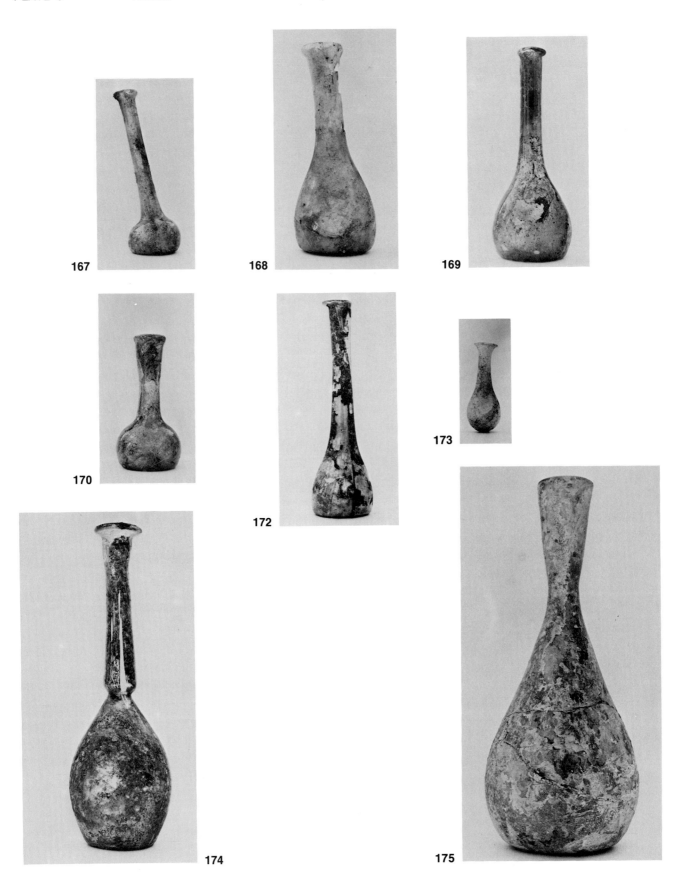

167

168

169

170

172

173

174

175

Jug with strap handles, ca. 2nd C. A.D.
Römisch-Germanisches Museum, Cologne
(435).

187

200

215

216

217 Impression.

233

PLATE 11   Early Byzantine

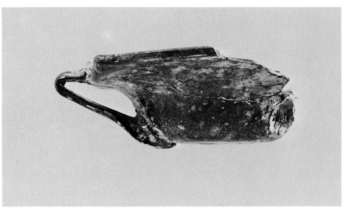

234

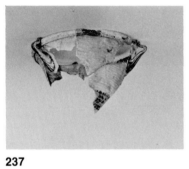

237

240

246

247

249

269

274

275

299

288

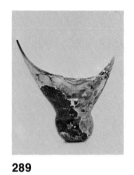

289

314

319

323

367

374

375

PLATE 13   Early Byzantine

382

391

395

414

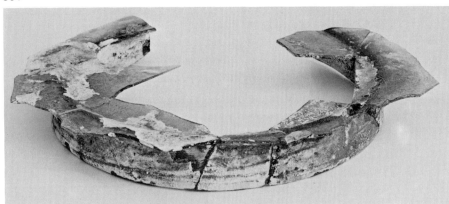

463

445

465

470

472

473

476

477

478

481

488

509

519

519 Showing diaphragm.

PLATE 15   Early Byzantine

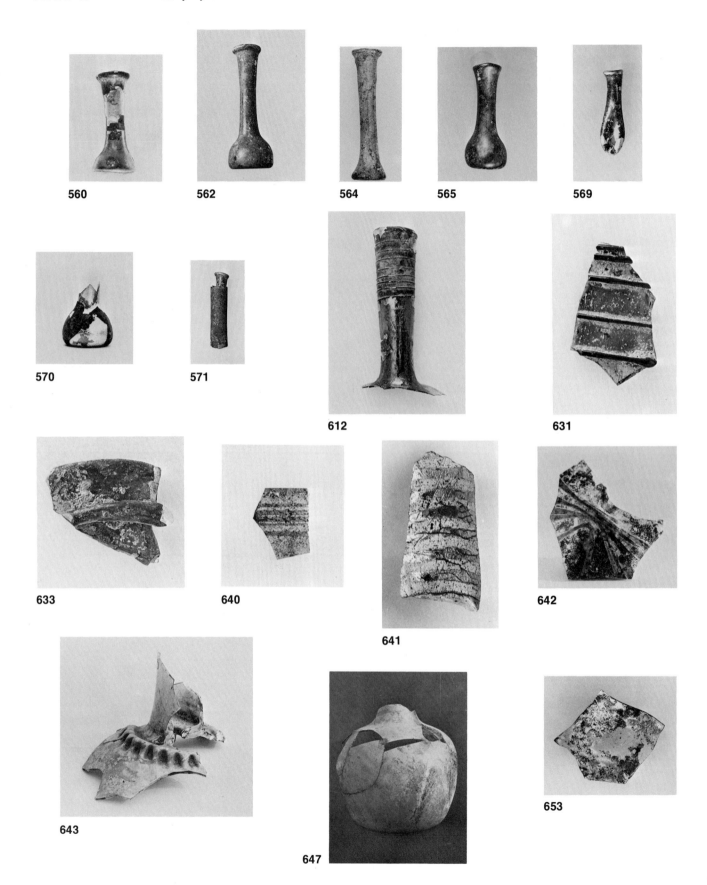

560

562

564

565

569

570

571

612

631

633

640

641

642

643

647

653

657

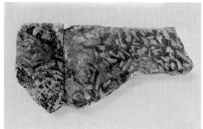

662

663

667

668

671

674

675   **676**   **677**

679

680–699 Selection of window pane fragments.

Fragment of window leading from BS E 6.

PLATE 17   Early Byzantine

707

Portion of **700** set in plaster.

722 Crucible coated with glass: cross section.

714

722 Int. glass residue.

725

729

733

722 Clay ext.

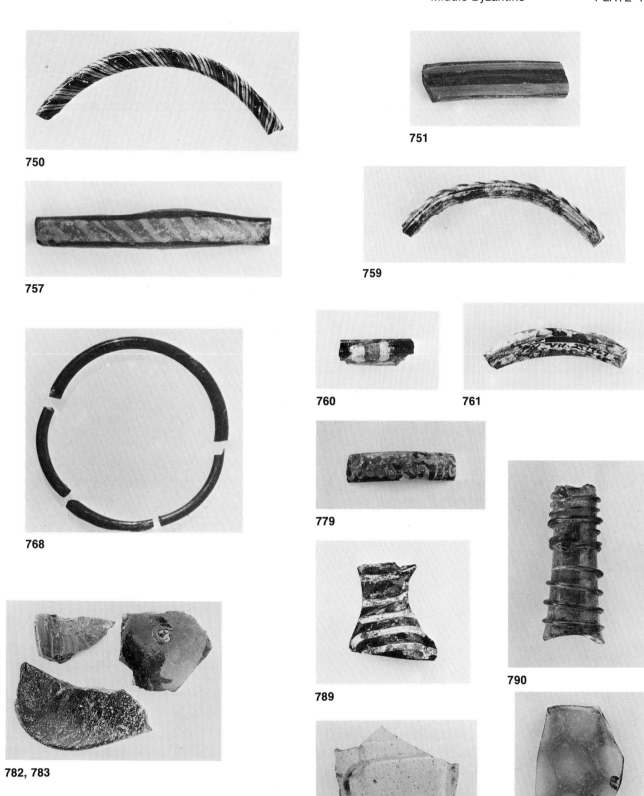

750

751

757

759

760

761

768

779

782, 783

789

790

791

794

PLATE 19   Ring Stones and Beads

817 (lower 1.) and sample ring stones.

821

824   **827**   **828**   **829**

830   **832**   **833**   **834**   **836**

837   **838**   **840**

845   **847**   **848**   **870**   **871**

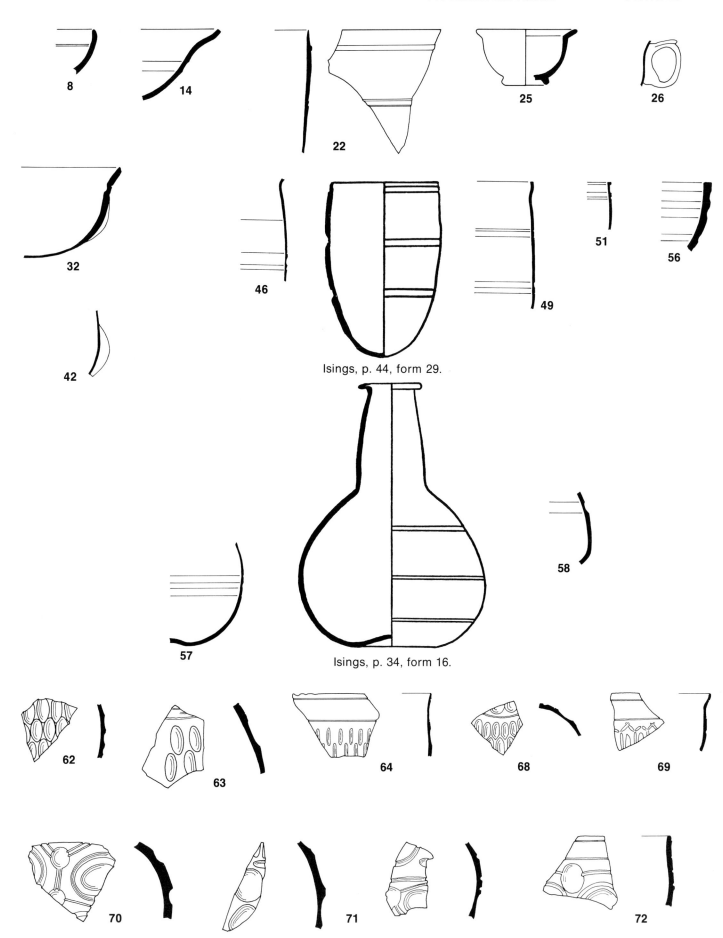

Isings, p. 44, form 29.

Isings, p. 34, form 16.

PLATE 21   Roman

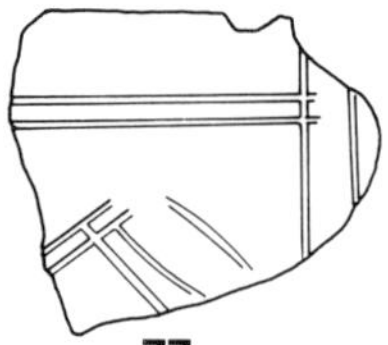

77

81

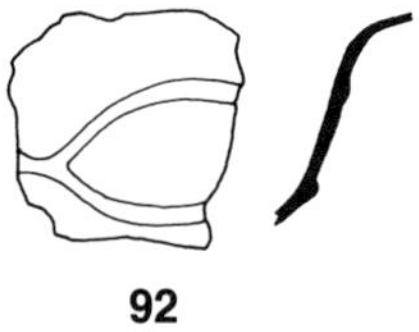

92

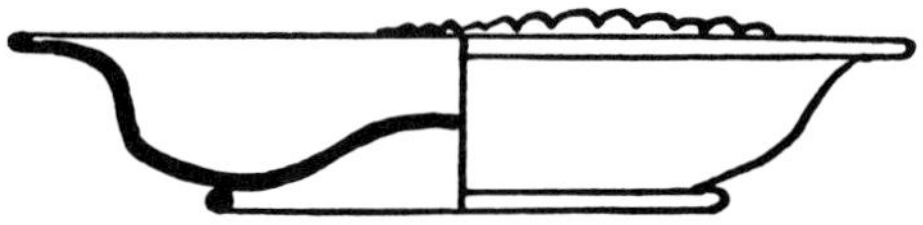

94

Isings, p. 59, form 43.

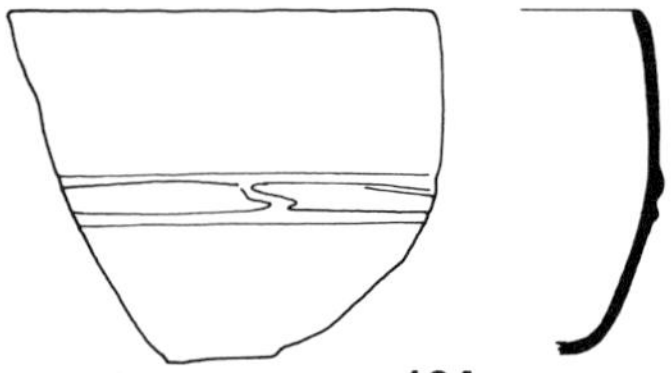

104

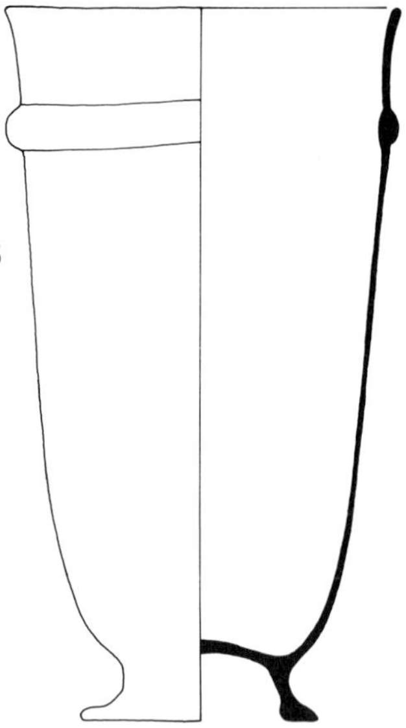

105

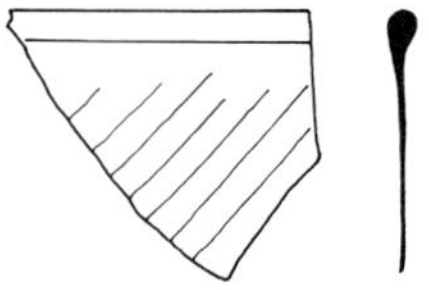

106

107

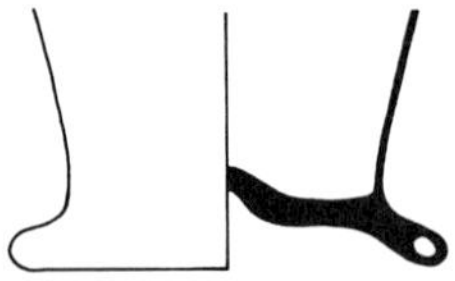

108

109

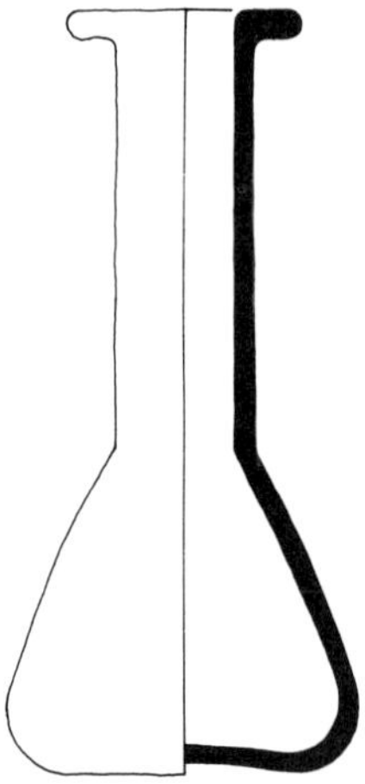

110

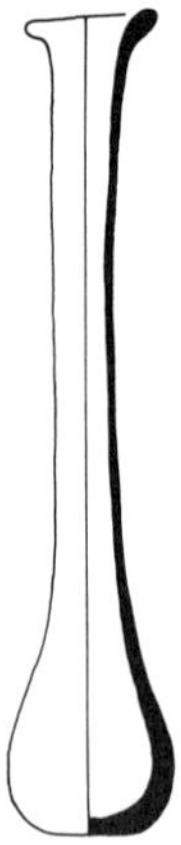

112

113

114

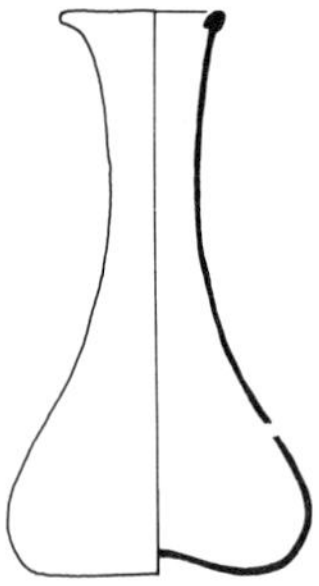

115

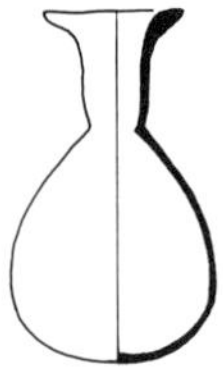

116

120

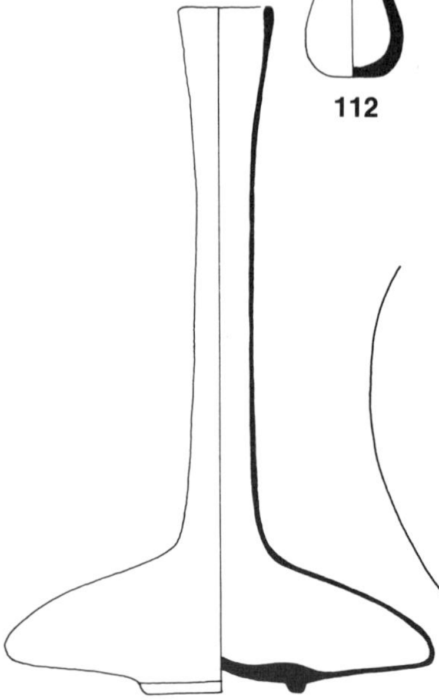

123

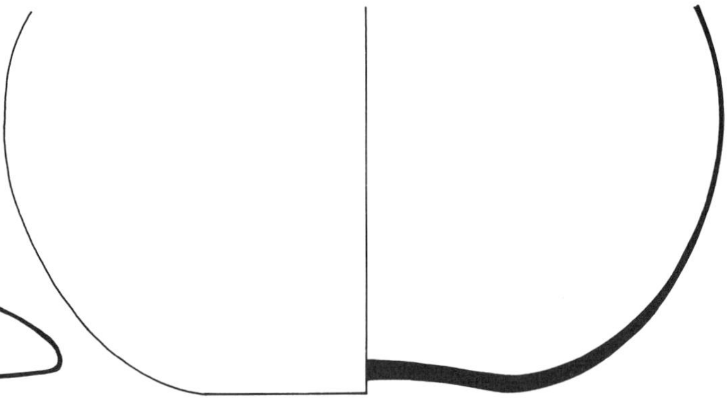

125

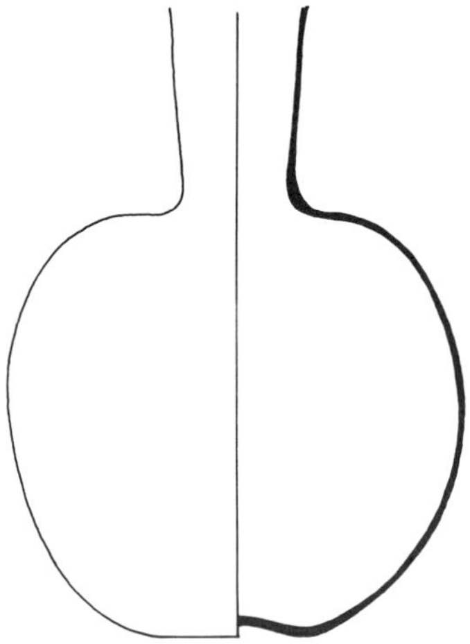

129

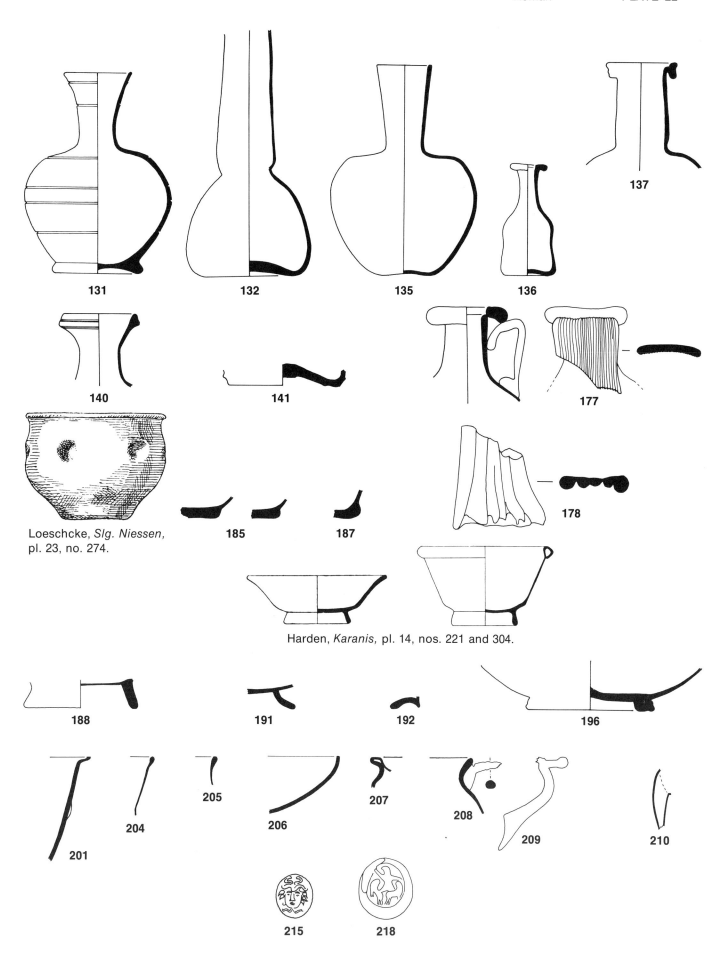

131   132   135   136   137

140   141   177

Loeschcke, *Slg. Niessen,* pl. 23, no. 274.

185   187   178

Harden, *Karanis,* pl. 14, nos. 221 and 304.

188   191   192   196

201   204   205   206   207   208   209   210

215   218

PLATE 23   Early Byzantine

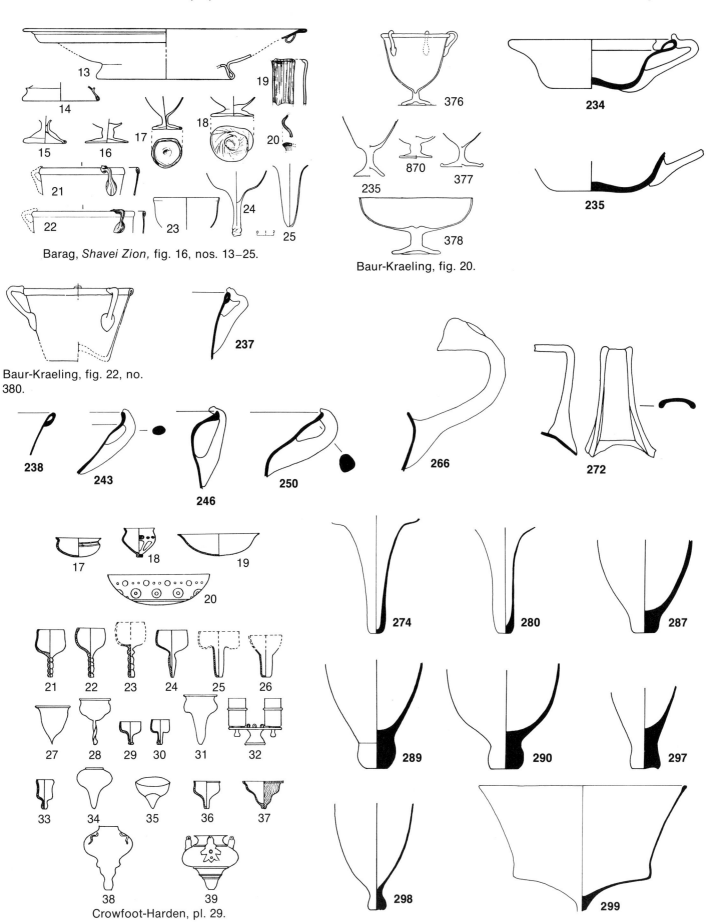

Barag, *Shavei Zion,* fig. 16, nos. 13–25.

Baur-Kraeling, fig. 20.

Baur-Kraeling, fig. 22, no. 380.

Crowfoot-Harden, pl. 29.

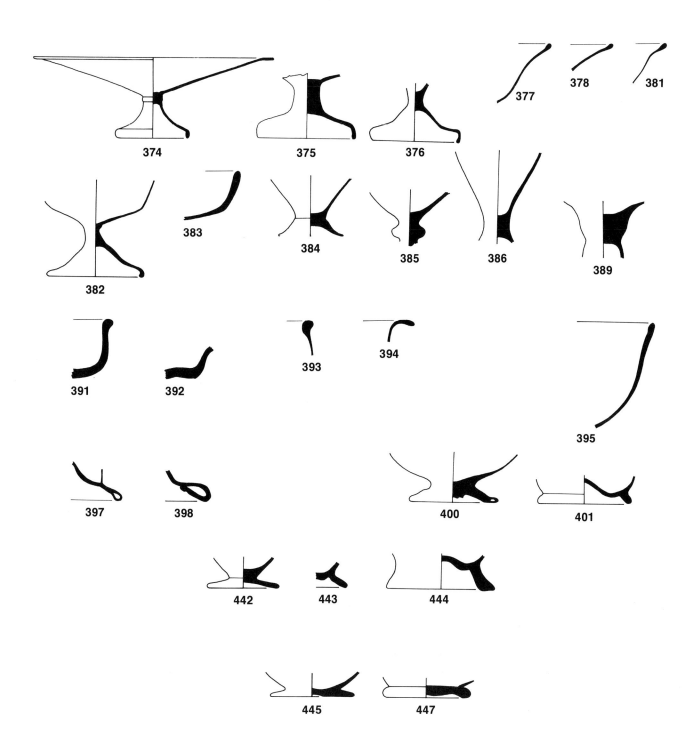

PLATE 25   Early Byzantine

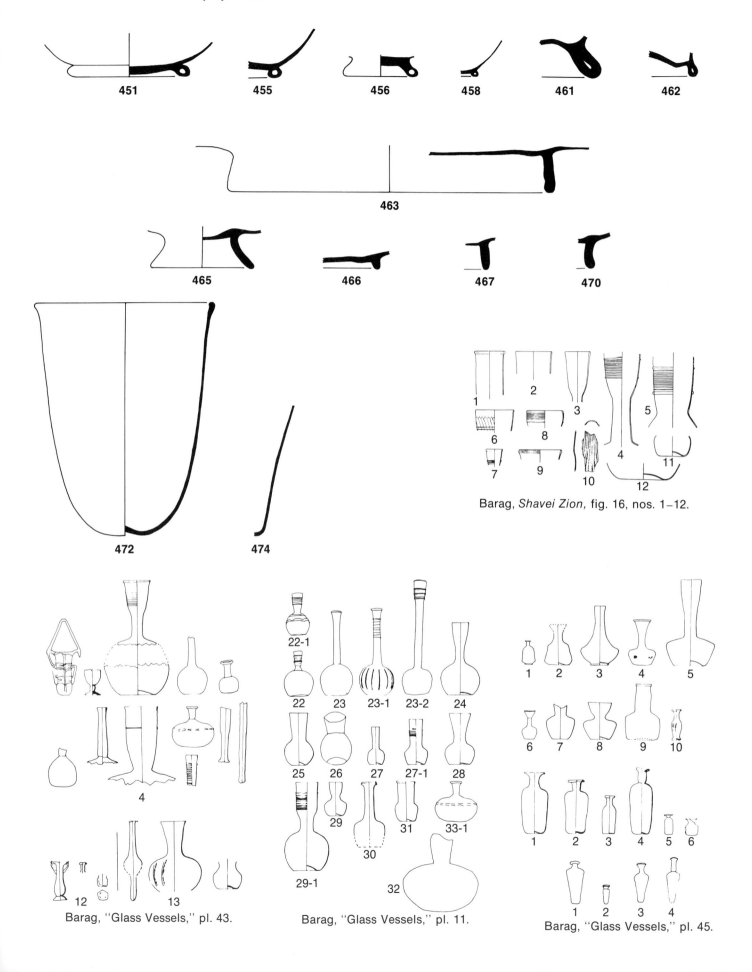

451   **455**   **456**   **458**   **461**   **462**

463

465   **466**   **467**   **470**

472   **474**

Barag, *Shavei Zion,* fig. 16, nos. 1–12.

Barag, "Glass Vessels," pl. 43.

Barag, "Glass Vessels," pl. 11.

Barag, "Glass Vessels," pl. 45.

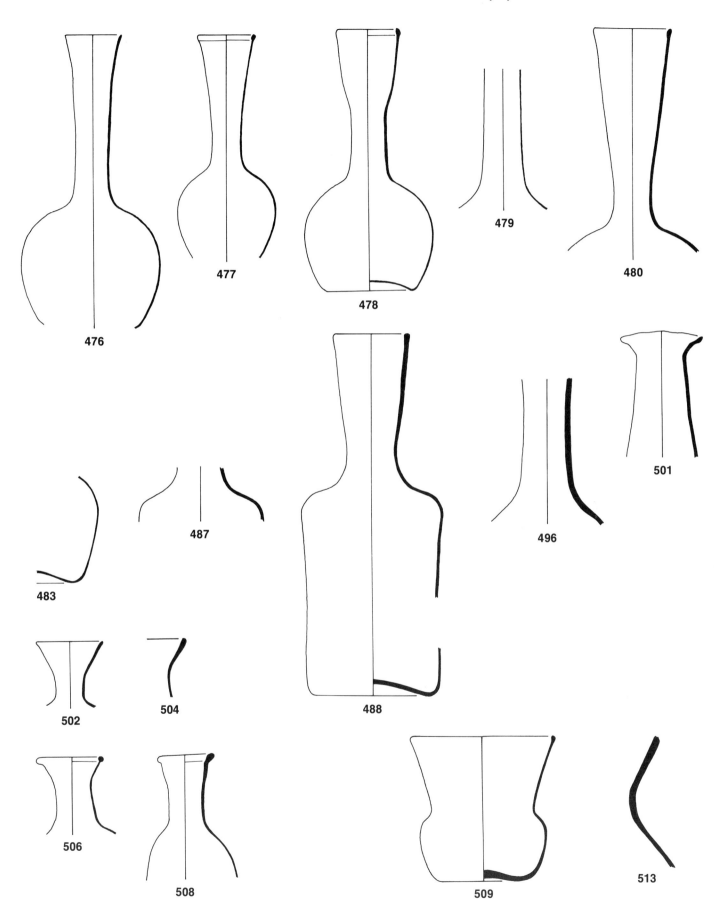

476

477

478

479

480

483

487

488

496

501

502

504

506

508

509

513

PLATE 27   Early Byzantine

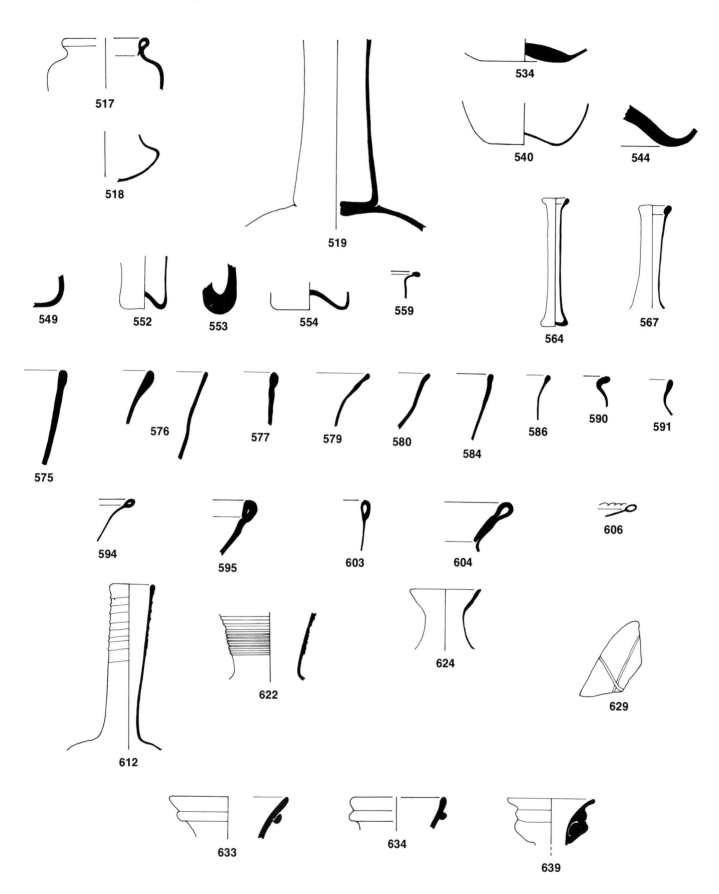

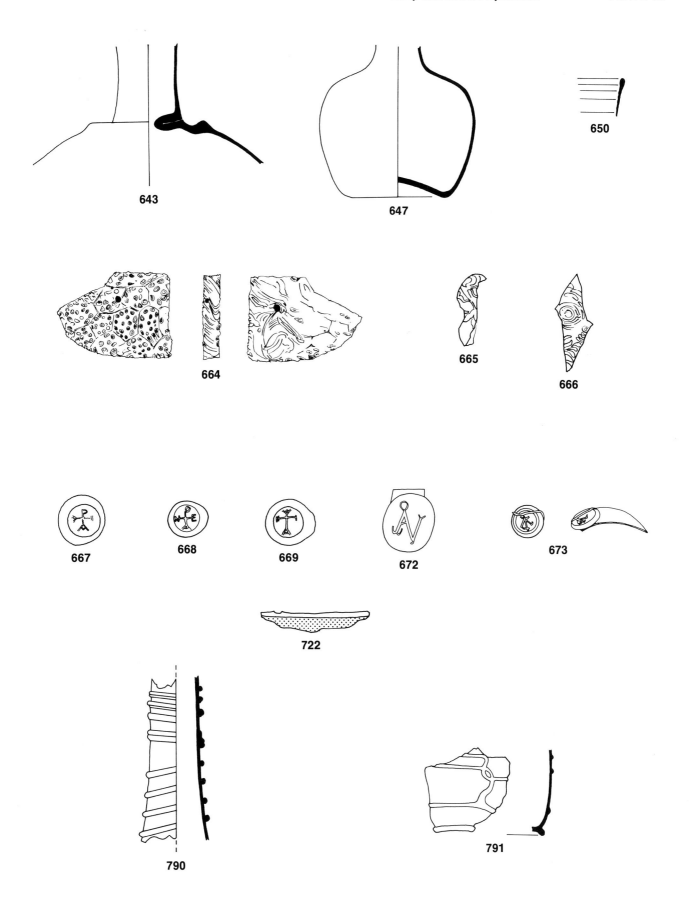

643

647

650

664

665

666

667

668

669

672

673

722

790

791